Space-Based Technologies in Poverty Alleviation for the Working Poor

Shelli Brunswick
SB Global, LLC, USA

Kyla L. Tennin
Center for Innovation and Education, USA

Tilottama Singh
Uttaranchal University, India

Richa Goel
https://orcid.org/0000-0002-8087-6568
Symbiosis International University, India

IGI Global
Scientific Publishing
Publishing Tomorrow's Research Today

Vice President of Editorial	Melissa Wagner
Director of Acquisitions	Mikaela Felty
Director of Book Development	Jocelynn Hessler
Production Manager	Mike Brehm
Cover Design	Jose Rosado

Published in the United States of America by
IGI Global Scientific Publishing
701 East Chocolate Avenue
Hershey, PA, 17033, USA
Tel: 717-533-8845 | Fax: 717-533-7115
Website: https://www.igi-global.com E-mail: cust@igi-global.com

Library of Congress Cataloging-in-Publication Data

LCCN: 2025044563 (CIP Data Pending)
ISBN13: 979-8-3373-7118-4
Isbn13Softcover: 979-8-3373-7119-1
EISBN13: 979-8-3373-7120-7
British Cataloguing in Publication Data
A Cataloguing in Publication record for this book is available from the British Library.

This book is dedicated to the working poor across the globe – individuals and families whose resilience, labor, and perseverance sustain economies yet too often go unseen, under-resourced, and undervalued. Your daily commitment to building a better future, even in the face of systemic inequalities, is both humbling and inspiring.

We also dedicate this work to the innovators, scientists, educators, and policy makers who believe that technology – especially space-based advancements – can and must be harnessed for the greater good. May this book serve as a bridge between the vastness of space and the grounded realities of human need, reminding us that progress is most meaningful when it is inclusive.

Finally, this work is dedicated to future generations – those who will inherit both the challenges and the possibilities of our world. May you continue to look upward, not only in exploration, but in pursuit of solutions that uplift humanity.

Table of Contents

Detailed Table of Contents

This chapter develops a sociological framework to examine how space-based technologies contribute to poverty alleviation among the working poor through the lens of spatial justice. While tools such as GIS, remote sensing, and satellite mapping enhance development planning, their application often remains technocratic, overlooking structural inequalities. Drawing on spatial justice, the capability approach, and social exclusion, the chapter argues that access to and control over spatial data are unevenly distributed, reflecting existing hierarchies. Using examples from India, it highlights how such technologies can either reinforce or reduce inequalities depending on inclusivity and governance. The chapter proposes a framework based on accessibility, participation, equity, and knowledge integration, advocating for a justice-oriented approach that centers the lived realities of marginalized populations.

This chapter examines the role of space technologies as potential tools for the development of the poor, especially the working poor, who continue to be excluded from the development process. It brings together theory and practice, to show how remote sensing, satellite communication, geospatial mapping and Earth observation technologies, can be integrated to improve agricultural productivity, increase access to education and healthcare, create jobs, and foster economic and social empowerment. The chapter also addresses the socio-technological, financial, and policy-related barriers to the use of the identified space technologies and suggests that these be implemented through multi-stakeholder partnerships, collaboration, and inclusive governance. From a high-tech to a low-tech perspective, this chapter views space technology as a strategy that fosters innovation and equity. It proposes that the linking of space technologies with terrestrial applications can help to diminish disparities and enhance sustainable development.

In this chapter, the geospatial technologies are discussed in terms of understanding and analysis of working poverty and socioeconomic inequality. Indeed, the spatial distribution of poverty and vulnerable populations can be visualized on the city and rural terrain by combining satellite imagery, remote sensing, and Geographic Information Systems (GIS) in order to locate vulnerable groups in both urban and rural environments. The chapter discusses relevant socioeconomic engines, which include the dynamic labor market, education, and access to infrastructure, and some of these spaces depict areas of inequality. Moreover, it gives the implications towards policymakers and planners set in evidence-based interventions and focused programs. Emerging technologies, including machine learning and mobile data integration, are also singled out as to be used to monitor poverty in real time, with a feeling of providing a means of making use of more inclusive and useful approaches to reduce poverty.

Traditional poverty data in LMICs rely on infrequent surveys, leaving gaps for timely decisions. Space technologies such as Earth observation and mobile data provide scalable, high-frequency proxies to produce granular poverty maps between censuses. Features like night-time lights, roads, buildings, and agriculture correlate with welfare, and when fused with surveys through "data sandwich" methods, improve precision and coverage. Machine learning, especially CNNs, boosts predictive accuracy and ensures consistent estimates across scales. Policy, Ethics and Future Directions in Alleviating Poverty from Space highlights alignment of EO with SDG mandates, robust data governance, and collaboration across sectors. Ethical use requires privacy-by-design, fairness audits, transparency, and community engagement. Evidence from Africa, Mexico, Bangladesh, Thailand, and Togo shows feasibility, accuracy, and integration into official statistics. Future directions stress multi-temporal monitoring, multimodal fusion, fairness-aware design, and EO-MLOps pipelines, as discussed in this chapter.

Working poverty, where individuals remain below poverty thresholds despite employment poses complex challenges that are deeply embedded in spatial and socio-economic contexts. Traditional statistical and econometric tools often struggle to capture the uncertainty, vagueness, and multidimensionality inherent in poverty indicators such as income stability, housing quality, healthcare access, and transport connectivity. This chapter introduces fuzzy soft sets as a novel mathematical framework for spatial poverty, the approach enables nuanced classification of poverty zones, flexible handling of incomplete or imprecise data, and multi-criteria decision support for resource allocation. Case illustrations demonstrate how fuzzy

Preface

INTRODUCTION

Poverty among the working poor represents one of the most persistent and complex challenges of our time. Unlike extreme poverty, the condition of being employed yet economically insecure often remains less visible, embedded within systems that mask vulnerability beneath participation in the labor force. This book emerges from a growing recognition that traditional approaches alone are insufficient to fully understand and address this form of poverty. New tools, new data, and new ways of thinking are required. Among the most promising of these are space-based technologies.

Advancements in satellite imagery, remote sensing, and global navigation systems have transformed how we observe and interpret the world. One confined to scientific exploration and national security, these technologies are now increasingly accessible and applicable to pressing social and economic challenges. They provide unprecedented capabilities to map, monitor, and analyze conditions on the ground – often in real time and at scales previously unimaginable. For populations that are difficult to reach through conventional data systems, space-based technologies offer a critical lens for visibility, inclusion, and informed intervention.

Space-Based Technologies in Poverty Alleviation for the Working Poor seeks to explore this intersection between technological innovation and social impact. The central premise of this volume is that poverty alleviation efforts can be significantly enhanced when satellite-derived data is integrated with socioeconomic indicators, local knowledge, and policy frameworks. By identifying poverty hotspots, monitoring environmental and economic risks, and informing targeted interventions, space-based approaches can contribute to more efficient, equitable, and sustainable development strategies.

This book brings together contributions from a diverse range of disciplines, reflecting the inherently interdisciplinary nature of the topic. Chapters examine advances in geospatial mapping, image processing, and domain adaptation, while also addressing critical considerations such as ethics, financial literacy, and socioeconomic inequality. The inclusion of topics such as renewable energy, green energy solutions, and medical imaging further underscores the broad applicability of space-based technologies in improving quality of life for vulnerable populations.

Importantly, this volume does not position technology as a standalone solution. Rather, it emphasized the importance of context – of aligning innovation with the lived realities, cultural dynamics, and specific needs of the working poor. The effectiveness of space-based interventions ultimately depends on how well they are integrated into local systems, governance structures, and community-driven initiatives. Ethical considerations, data equity, and accessibility remain central to ensuring that technological advancements do not inadvertently widen the very gaps they aim to close.

This work is intended for researchers, academicians, and policymakers, development practitioners, and students who are interested in exploring new frontiers in poverty alleviation. It aims to serve both as a scholarly resource and as a practical guide for those seeking to leverage space-based technologies for inclusive development. By bridging technical knowledge with social insight, the book aspires to foster dialogue, inspire innovation, and support evidence-based decision-making.

As the world continues to confront economic uncertainty, climate change, and widening inequalitites, the need for scalable and forward-looking solutions becomes increasingly urgent. Space-based technologies, when used thoughtfully and responsibly, hold immense potential to contribute to this effort. It is our hope that this volume not only advances academic discourse but also catalyzes meaningful action toward improving the lives and livelihoods of the working poor worldwide.

The persistence of poverty among the working poor presents a paradox that continues to challenge scholars, policymakers, and practitioners alike. Employment has long been regarded as a primary pathway out of poverty; however, for millions of individuals around the world, work does not guarantee economic security or upward mobility. Instead, many remain caught in cycles of low wages, limited benefits, and vulnerability to external shocks such as environmental hazards, health crises, and market fluctuations. This nuanced form of poverty often escapes traditional measurement systems, making it more difficult to diagnose and address effectively. In response, there is an increasing need for innovative approaches that can capture both the visible and hidden dimensions of economic hardship. Space-based technologies offer one such approach, providing new avenues for understanding and responding to the realities faced by the working poor.

Over the past several decades, the evolution of satellite systems, geospatial analytics, and remote sensing capabilities has significantly expanded the scope of data available to researchers and decision-makers. These technologies make it possible to observe patterns of human settlement, infrastructure development, environmental change, and resource distribution with remarkable accuracy. More importantly, they enable the integration of spatial data with socioeconomic indicators, creating a more comprehensive and dynamic picture of poverty. In regions where data collection is limited or inconsistent, satellite-derived insights can fill critical gaps, offering a level of visibility that was previously unattainable.

This volume, *Space-Based Technologies in Poverty Alleviation for the Working Poor*, is grounded in the belief that technological innovation must be purposefully aligned with social impact. It examines how space-based tools can be applied not only to identify areas of need but also to inform targeted, data-driven interventions. From mapping underserved communities to supporting disaster preparedness and response, these technologies have the potential to enhance both the efficiency and effectiveness of poverty alleviation efforts. By combining technical methodologies with contextual understanding, the book highlights how data can be transformed into actionable knowledge that benefits vulnerable populations.

The contributions within this book reflect a wide range of academic disciplines and practical perspectives. Topics such as geospatial mapping, image processing, and domain adaptation are explored alongside critical discussions of ethics, financial inclusion, and inequality. This interdisciplinary approach underscores the complexity of poverty as a global issue and the necessity of drawing on multiple fields of expertise to address it. Additionally, the inclusion of areas such as renewable energy, green technologies, and medical imaging demonstrates the far-reaching implications of space-based applications in improving living conditions and expanding access to essential services.

A central theme throughout this work is the recognition that technology alone cannot resolve deeply rooted social challenges. The successful application of space-based solutions depends on their integration into broader systems that include governance, community engagement, and cultural awareness. Local knowledge and participation are essential to ensuring that interventions are relevant, equitable, and sustainable. At the same time, ethical considerations must remain at the forefront, particularly in relation to data use, privacy, and representation. As the availability of geospatial data continues to grow, so too does the responsibility to use it in ways that promote fairness and avoid unintended consequences.

The book also considers the evolving landscape of global collaboration in the use of space-based technologies. Increasingly, partnerships between governments, private sector entities, academic institutions, and international organizations are shaping how these tools are developed and deployed. Open data initiatives and shared platforms are making it easier for a broader range of stakeholders to access and utilize satellite information. This shift toward greater accessibility has the potential to empower local actors and foster innovation at multiple levels. However, it also highlights the importance of building capacity and ensuring that users have the skills and resources needed to interpret and apply complex data effectively.

Intended for a diverse audience that includes researchers, educators, policymakers, and development professionals, this book aims to contribute to both academic inquiry and practical application. It serves as a resource for those seeking to understand the intersection of space technology and social development, while also offering insights for those actively engaged in designing and implementing poverty reduction strategies. By presenting both theoretical frameworks and applied examples, the volume encourages readers to think critically about how emerging technologies can be leveraged to create more inclusive and resilient societies.

In an era marked by rapid technological advancement and persistent inequality, the need for innovative, scalable solutions has never been greater. Space-based technologies represent a unique and evolving frontier in this effort, offering tools that can enhance our ability to see, understand, and respond to complex global challenges. While they are not panacea, their thoughtful application can play a meaningful role in advancing equitable development and improving the livelihoods of the working poor. It is our hope that this book inspires continued exploration, collaboration, and action in pursuit of a more just and sustainable world.

CHAPTER OVERVIEW

Chapter 1. Spatial Justice and the Working Poor: A Sociological Framework for Space-Based Poverty Alleviation establishes a sociological foundation for understanding the role of space-based technologies in addressing poverty among working populations. The chapter examines how Geographic Information Systems (GIS), remote sensing, and satellite mapping can support development planning while also potentially reproducing existing social and spatial inequalities when access to and control over data remain uneven. Drawing on spatial justice, the capability approach, and social exclusion, it emphasizes the importance of accessibility, participation, equity, and knowledge integration. Through examples from India, the chapter demonstrates how inclusive governance and attention to the lived experiences of marginalized communities can transform space technologies from predominantly technical instruments into tools for advancing social justice and equitable poverty alleviation.

Chapter 2. From Satellites to Solutions: Leveraging Space Technologies to Uplift the Working Poor examines how space technologies can be translated into practical development interventions for populations that remain excluded from conventional development processes. The chapter brings together remote sensing, satellite communication, geospatial mapping, and Earth observation to demonstrate potential applications in agricultural productivity, education, healthcare, employment, and broader economic and social empowerment. Alongside these opportunities, it considers technological, financial, and policy barriers that can limit adoption, emphasizing the importance of multi-stakeholder partnerships, collaboration, and inclusive governance. By connecting high-tech space-based capabilities with accessible terrestrial applications, the chapter presents space technology as a means of fostering innovation, reducing disparities, and advancing more equitable and sustainable development.

Chapter 3. Geospatial Mapping of Working Poverty and Socioeconomic Inequality examines the use of geospatial technologies to identify and analyze the spatial dimensions of working poverty and socioeconomic inequality across urban and rural settings. By integrating satellite imagery, remote sensing, and GIS, the chapter demonstrates how vulnerable populations and patterns of deprivation can be visualized geographically and linked to socioeconomic factors such as labor-market conditions, education, and infrastructure access. It highlights the value of spatially differentiated evidence for enabling policymakers and planners to design targeted interventions and more effectively allocate resources. The chapter also considers emerging applications of machine learning and mobile data integration, which offer opportunities for more timely poverty monitoring and the development of increasingly inclusive, responsive, and evidence-based approaches to poverty reduction.

Chapter 4. Policy, Ethics and Future Directions in Alleviating Poverty from Space addresses the potential of Earth observation and mobile data to overcome limitations associated with conventional poverty measurement, particularly in low- and middle-income countries where surveys may be infrequent and unable to provide sufficiently granular information. The chapter explores the use of indicators such as nighttime lights, roads, buildings, and agricultural features, combined with survey information through data-sandwich approaches, to generate more frequent and detailed poverty estimates. It further examines the contribution of machine learning techniques to improving predictive accuracy and supporting consistent estimates across geographic scales. Beyond technological capabilities, the chapter places strong emphasis on policy alignment, data governance, privacy, fairness, transparency, and community engagement. Drawing on evidence from multiple international contexts, it outlines future directions involving multi-temporal monitoring, multimodal data fusion, fairness-aware systems, and Earth-observation machine-learning pipelines.

Chapter 5. Spatial-Fuzzy Integration for Poverty Response: Mapping Uncertainty, Vulnerability, and Policy Action introduces a mathematical and computational perspective for addressing the uncertainty and multidimensionality inherent in the measurement of working poverty. Recognizing that conventional statistical approaches may have difficulty representing imprecise or overlapping conditions involving income stability, housing, healthcare, and transportation access, the chapter presents fuzzy soft sets as a framework for spatial poverty analysis. This approach enables more nuanced classifications of poverty zones, accommodates incomplete or uncertain data, and supports multi-criteria decision-making for resource allocation. Through illustrative applications, the chapter demonstrates how fuzzy soft sets can identify overlapping vulnerability clusters, prioritize districts for intervention, and enhance spatial decision-support systems. In doing so, it expands the methodological foundations of poverty research while encouraging connections among mathematical modeling, social science, geospatial analysis, and public policy.

Chapter 6. Mapping Working Poverty from Space: A Geospatial Approach to Socioeconomic Inequality further explores the transformative potential of geospatial intelligence for understanding the distribution and complexity of working poverty. By combining satellite imagery, remote sensing indicators, GIS-based analysis, and socioeconomic information, the chapter demonstrates how spatial technologies can reveal patterns of deprivation, informal-sector concentration, and unequal access to essential services that may remain difficult to identify through conventional approaches. Attention is given to the challenges of mapping poverty at fine spatial scales and to the interactions among environmental, demographic, and economic conditions that shape vulnerability. Through analytical frameworks and case-based discussion, the chapter considers how spatial intelligence can support data-driven interventions, strengthen social protection systems, and inform more inclusive development planning, while also addressing the ethical considerations associated with the collection and application of spatial data.

Shelli Brunswick

SB Global, LLC, USA

Kyla L. Tennin

Center for Innovation and Education, USA

Tilottama Singh

Uttaranchal University, India

Richa Goel

Symbiosis International University, India

Acknowledgement

The development of *Space-Based Technologies in Poverty Alleviation for the Working Poor* has been a deeply collaborative and interdisciplinary effort, and we extend our sincere gratitude to all who contributed to its realization.

We are especially thankful to the researchers, scholars, and practitioners whose work in geospatial mapping, remote sensing, socioeconomic analysis, and sustainable development has laid the intellectual foundation for this volume. Your commitment to advancing knowledge at the intersection of technology and social impact continues to shape meaningful global change.

We acknowledge the contributions of policy makers and development organizations who are actively working to integrate space-based technologies into poverty alleviation strategies. Your efforts to translate innovation into action demonstrate the transformative potential of data-driven, inclusive policy design.

We also extend our appreciation to the academic and professional communities whose insights into ethics, financial literacy, renewable energy, and socio-economic inequality have enriched the scope and depth of this work. The integration of diverse perspectives has strengthened the interdisciplinary nature of this publication.

Special thanks are due to the reviewers and editorial team whose thoughtful feedback, guidance, and dedication ensured the rigor and clarity of this book. Your commitment to excellence has been invaluable.

Lastly, we express our gratitude to the communities whose lived experiences continue to inform and inspire this work. It is our hope that this book contributes, in some meaningful way, to advancing equitable opportunities and improving livelihoods for the working poor worldwide.

Introduction

The challenge of poverty among the working poor stands at the intersection of economic participation and structural inequality. Across both developed and developing contexts, millions of individuals engage in consistent labor yet remain unable to achieve financial stability or long-term security. This condition reflects not only wage insufficiencies but also systematic limitations in access to resources, infrastructure, and opportunities. Traditional methods of poverty assessment – often reliant on household surveys, census data, and localized reporting – have provided valuable insights, yet they frequently fall short in capturing the dynamic, spatial, and often hidden dimensions of working poverty. In this evolving landscape, space-based technologies offer a transformative lens through which poverty can be more accurately identified, understood, and addressed.

Satellite imagery, remote sensing, and global navigation systems have expanded the boundaries of data collection and analysis. These technologies enable continuous, large-scale observation of environmental, economic, and infrastructural conditions, allowing for a level of precision and timeliness that ground-based systems alone cannot achieve. From detecting informal settlements and monitoring agricultural productivity to assessing disaster impacts and infrastructure gaps, space-based tools provide critical inputs for understanding the lived realities of economically vulnerable populations. When combined with socioeconomic datasets, these technologies create a multidimensional view of poverty – one that is both geographically grounded and analytically robust.

Space-Based Technologies in Poverty Alleviation for the Working Poor is positioned within this emerging paradigm. The book seeks to bridge the gap between technological capability and social application by examining how space-based innovations can be leveraged to support inclusive development. It brings together interdisciplinary perspectives that span space engineering, geospatial analytics, social sciences, and public policy, offering a comprehensive exploration of how data derived from space can inform targeted interventions for the working poor.

A central premise of this volume is that poverty is not only an economic condition but also a spatial phenomenon. Where people live, work, and interact with their environment significantly influences their access to opportunities and resources. As such, geospatial mapping and spatial analysis become essential tools in identifying poverty hotspots and understanding the distribution of inequality. By applying a spatial lens, researchers and practitioners can uncover patterns that may otherwise remain obscured, enabling more strategic allocation of resources and more effective program design.

To guide this exploration, the book introduces two conceptual frameworks that underpin the integration of space-based technologies into poverty alleviation strategies.

FRAMEWORK 1: THE SPATIAL POVERTY INTELLIGENCE FRAMEWORK (SPIF)

The Spatial Poverty Intelligence Framework provides a structured approach to identifying, analyzing, and responding to working poverty through geospatial data integration. This framework consists of four interconnected components:

1. **Data Acquisition and Integration –** Leveraging satellite imagery, remote sensing data, and global navigation systems alongside traditional socioeconomic indicators such as income levels, employment patterns, and access to services.
2. **Spatial Analysis and Mapping –** Utilizing geospatial tools and image processing techniques to identify patterns of deprivation, infrastructure gaps, and environmental risks that disproportionately affect the working poor.

3. **Insight Generation and Interpretation** – Applying domain adaptation and advanced analytics to translate raw data into meaningful insights that reflect both macro-level trends and localized conditions.
4. **Targeted Intervention Design** – Informing policy decisions, development programs, and resource allocation strategies based on spatially informed evidence, ensuring interventions are both precise and contextually relevant.

Table 1. A representation of framework #1- Spatial Poverty Intelligence Framework (SPIF)

Framework Component	Description	Key Tools/Methods	Expected Outcomes
1. Data Acquisition & Integration	Collection and combination of satellite data with socioeconomic indicators	Satellite imagery, remote sensing, census data, surveys	Comprehensive, multi-source dataset capturing economic and spatial conditions
2. Spatial Analysis & Mapping	Identification of geographic patterns of poverty and infrastructure gaps	GIS mapping, geospatial analytics, image processing	Visualized poverty hotspots and spatial distribution of inequality
3. Insight Generation & Interpretation	Transformation of raw data into meaningful, actionable insights	Machine learning, domain adaptation, data modeling	Contextual understanding of poverty drivers and localized vulnerabilities
4. Targeted Intervention Design	Development of policies and programs informed by spatial intelligence	Policy modeling, decision-support systems, stakeholder input	Precision-based interventions tailored to community needs and environmental conditions

The SPIF emphasizes the importance of integrating multiple data sources and analytical methods to create a holistic understanding of poverty. It also highlights the role of continuous monitoring, enabling stakeholders to assess the impact of interventions over time and adapt strategies as needed.

FRAMEWORK 2: THE INCLUSIVE SPACE TECHNOLOGY APPLICATION MODEL (ISTAM)

While the SPIF focuses on data and analysis, the Inclusive Space Technology Application Model centers on implementation and impact. This framework outlines how space-based technologies can be effectively translated into real-world solutions that benefit the working poor. It includes five key dimensions:

1. **Accessibility** – Ensuring that space-based data and tools are available to a wide range of stakeholders, including local governments, community organizations, and small enterprises.
2. **Affordability** – Promoting cost-effective solutions that can be scaled and sustained, particularly in resource-constrained settings.
3. **Relevance** – Aligning technological applications with the specific needs and priorities of the working poor, including areas such as financial inclusion, disaster resilience, and access to energy.
4. **Ethical Governance** – Addressing issues related to data privacy, equity, and representation, and ensuring that the use of space-based technologies adheres to principles of fairness and accountability.
5. **Capacity Building** – Supporting the development of skills, knowledge, and institutional frameworks to effectively utilize and manage space-based tools.

Figure 1. An inclusive space technology model diagram

Note. A visual representation for framework number two, titled "The Inclusive Space Technology Application Model (ISTAM)".

Together, these frameworks provide a comprehensive foundation for understanding both the potential and the limitations of space-based technologies in poverty alleviation. They underscore the importance of not only generating high-quality data but also ensuring that such data is translated into actionable, equitable, and sustainable outcomes.

The chapters that follow build upon these frameworks by exploring a wide array of topics, including domain adaptation techniques for improving data accuracy, ethical considerations in geospatial analysis, and the role of financial literacy in leveraging technological insights for economic empowerment. Discussions on renewable and green energy highlight how satellite data can support sustainable infrastructure development, while explorations of medical imaging and health mapping demonstrate the broader applications of space-based technologies in enhancing human well-being.

This book is intended for a diverse audience, including researchers, academicians, policymakers, sociologists, and practitioners engaged in development work. It aims to serve as both a theoretical guide and a practical resource, offering insights that can inform research, policy, and implementation. By fostering interdisciplinary dialogue and encouraging innovative thinking, the volume seeks to contribute to a growing body of knowledge that positions technology as a catalyst for social good.

As global challenges such as climate change, urbanization, and economic inequality continue to evolve, the need for integrated, data-driven solutions becomes increasingly urgent. Space-based technologies, when applied thoughtfully and inclusively, hold significant promise in addressing these challenges. This introduction sets the stage for a deeper exploration of how such technologies can be harnessed to improve the lives and livelihoods of the working poor, ultimately contributing to a more equitable and sustainable future.

Kyla L. Tennin

Center for Innovation and Education, USA

Chapter 1
Spatial Justice and the Working Poor: A Sociological Framework for Space-Based Poverty Alleviation

Kritika Sharma
http://orcid.org/0009-0001-2884-557X
Lovely Professional University, Jalandher, India

Harry Kaur
http://orcid.org/0000-0001-8276-2313
Lovely Professional University, Jalandher, India

ABSTRACT

This chapter develops a sociological framework to examine how space-based technologies contribute to poverty alleviation among the working poor through the lens of spatial justice. While tools such as GIS, remote sensing, and satellite mapping enhance development planning, their application often remains technocratic, overlooking structural inequalities. Drawing on spatial justice, the capability approach, and social exclusion, the chapter argues that access to and control over spatial data are unevenly distributed, reflecting existing hierarchies. Using examples from India, it highlights how such technologies can either reinforce or reduce inequalities depending on inclusivity and governance. The chapter proposes a

DOI: 10.4018/979-8-3373-7118-4.ch001

framework based on accessibility, participation, equity, and knowledge integration, advocating for a justice-oriented approach that centers the lived realities of marginalized populations.

INTRODUCTION

The persistence of poverty among the working poor has always been the main issue in modern development discussions, especially in the Global South, where economic growth has not been accompanied by equal distribution of resources and opportunities. The working poor are individuals who work yet are unable to attain the minimum standard of living despite their labor, revealing a paradox that undermines the traditional view of employment as a means of escaping poverty (Van der Linden, 2019). This condition is highly complex, as it is closely interrelated with structural inequalities shaped by caste, class, gender, and region, which lead to uneven access to livelihoods, infrastructure, and social protection systems (De Siqueira, 2012). These disparities are increasingly geographic and social as poverty is concentrated in geographically marginalized areas, such as rural areas, informal urban slums, and marginalized communities.

Simultaneously with these problems, the past twenty years have seen a swift surge in more sophisticated technologies in development planning and governance, including the deployment of artificial intelligence (AI), machine learning, and space-based technologies such as satellite imagery, Geographic Information Systems (GIS), and remote sensing. The ability to generate big data, provide predictive analytics, and enhance the accuracy of policy interventions is frequently praised as a key feature of these technologies (Citaristi, 2022). For example, satellite-based surveillance has been used to measure agricultural productivity, forecast climatic hazards, map disaster-prone areas, and target underserved populations. On the same note, AI-based data analytics are increasingly being utilized to streamline welfare targeting and resource allocation (Melnyk et al., 2020). Combined, these innovations indicate a transition to a data-driven form of governance, with technological efficiency serving as a solution to complex socio-economic issues.

Yet, this increased dependency on technology has some very important sociological issues. Although AI and space-based tools can improve the ability to observe and quantify poverty, they do not always focus on the structural conditions that create and perpetuate it. Researchers have claimed that data-centric practices tend to depoliticize poverty by treating it as an issue of information gaps rather than a result of historical and systemic inequalities (Kitchin, 2021). Furthermore, technological infrastructure, digital literacy, and data representation are also unevenly distributed, contributing to the emergence of new forms of exclusion commonly referred to as the digital divide (Moravec, 2024). Consequently, they may make marginalized groups, such as the working poor, visible in data sets, but they are not heard in the decision-making process. This contradiction underscores the importance of questioning the social consequences of technological interventions in development twice over.

In this respect, the concept of spatial justice becomes an important tool of analysis. The concept of spatial justice has been put forward as follows (Iveson, 2011): it argues that social inequalities are geographic and that spatial organization is a central factor in determining access to resources, opportunities, and power. It highlights the role of uneven development, infrastructural inequalities, and territorial governance in generating and reproducing patterns of advantage and disadvantage in regions. In this sense, poverty is not only a state of income deprivation but also a geographically determined feature related to the inability to participate in systems of connectivity, services, and institutional assistance. The incorporation of space-based technologies into development processes, consequently, needs to be assessed not only in terms of efficiency but also in terms of equity and justice.

In this chapter, an attempt has been made to fill this gap by formulating a sociological conceptualization of space-based poverty alleviation that centers on the experiences and realities of the working poor. It relies on theoretical viewpoints, such as spatial justice, the capability approach (Gangas, 2016), and social exclusion theories, to critically analyses how technologies that operate in space may help reduce and create inequalities. The chapter specifically lays stress on the Indian experience in which regional inequalities, rural marginalization, and the existence of indigenous and marginalized communities offer a fertile ground in considering the intersections of space, technology and poverty.

This chapter has fourfold objectives. First, it attempts to theorize the connection between spatial inequality and poverty, as those geographic entities create the lived experiences of the working poor. Second, it aims to analyze how space-based and AI-driven technologies can be applied in the process of developmental intervention, and their possibilities and constraints. Third, it examines how these technologies overlap with existing systems of exclusion, such as issues of access, representation, and participation. Lastly, the chapter proposes a sociological model that integrates technological innovation with principles of equity, inclusivity, and local knowledge systems.

The significance of the study lies in its interdisciplinary methodology, which draws on sociology, development studies, and the analysis of technologies. The chapter shifts the focus to technocracy by preempting the idea of spatial justice and highlighting the necessity of socially oriented, ethically conscious methods of poverty reduction. In this way, it provides a more detailed insight into how emerging technologies could be used not only to be efficient, but also to promote justice and inclusion in development processes.

LITERATURE REVIEW

Over the last several years, there has been a growing scholarly interest in the convergence of poverty, technology, and spatial inequality. Nevertheless, the current research is still scattered across fields, and there is a weak connection between the sociological theory and the application of technology. In this section, the review examines major strands of scholarship pertinent to the working poor, spatial inequality, and the use of space-based and AI-driven technologies in development, identifying critical gaps that this chapter aims to fill.

Poverty and the Working Poor: Beyond Income-Based Approaches

Historical approaches to poverty have focused primarily on income deprivation and economic measures, and in most cases have paid little attention to the multidimensionality of poverty. Early developmental theories emphasized economic development as the main channel for poverty alleviation, since the benefits would trickle down to marginalized groups. Nevertheless, the assumption has always been challenged by empirical evidence, especially for working poor people who remain poor even when working (Kornbluh, 2009).

The move toward multidimensional conceptualizations of poverty (primarily the capability approach) has greatly broadened the analytical prism. According to (Wagle, 2002) poverty is best comprehended as a denial of the capabilities that the real freedoms available to people are to live the kind of lives they appreciate instead of just being deprived of income. This school of thought emphasises the need to have access to education, healthcare, infrastructure, and social participation. Based on this, (Alkire & Santos, 2013) came up with multidimensional poverty indices that reflect non-monetary information, which further reiterates the intricacy of deprivation.

Another aspect of poverty that sociological studies have been trying to focus on is the structural aspect of poverty, associating it with inequality systems like caste, class, gender and ethnicity. Regarding the Indian scenario, researchers have demonstrated that marginalized groups, including marginalized community populations and informal laborers, are continuously sidelined in the developmental process (Desai et al., 2025). These analyses indicate that poverty is not merely financial but has deep roots in social networks and social institutions.

Spatial Inequality and the Concept of Spatial Justice

Much literature has highlighted the spatial aspects of poverty and inequality. The very idea of spatial justice (Israel & Frenkel, 2018) builds on earlier work by Lefebvre (1991) and Harvey (2008), who suggest that space is a socially constructed phenomenon and that inequalities are, in any way, geographic. The physical design of the space, such as urban design, infrastructure distribution, and land use, is a significant determinant of access to resources and opportunities.

Empirical evidence shows how spatial inequality is experienced across rural and urban settings. Geographic isolation, connectivity, and market and service accessibility in rural settings are factors that can lead to persistent poverty (Constantin, 2021). Informal settlements and slums grow unevenly across the city, indicating both underdevelopment and non-integration into planning procedures (Uwayezu & de Vries, 2019). Such spatial inequalities are neither accidental nor incidental but are usually influenced by historical and political events that favour some geographical areas and citizens more than others.

The impact of spatial injustice to strengthen social exclusion has also been put forward by scholars. Indicatively, the lack of infrastructure in marginalized neighborhoods restricts access to education, health institutions, and job opportunities, thereby constituting a continuous experience of poverty (Harvey, 2010). In this respect, poverty can be addressed through economic solutions and redistribution of space, thereby making resource distribution more equal.

Space-Based Technologies in Development

There has been a significant increase in the application of space-based technologies in development, including agriculture, disaster management, urban planning, and environmental monitoring. Governments and international organizations have embraced Geographic Information Systems (GIS), remote sensing, and satellite imagery as evidence-based decision-making tools (Harvey, 2010).

Satellite data has been applied in agriculture to monitor crop health, forecast harvests, and manage resources to improve productivity and resilience (FAO). Remote sensing technologies in disaster management also enable real-time observation of floods, droughts, and other risks, enabling early warning systems and targeted interventions (Villagran de Leon, 2023). On the same note, geospatial technologies are used in urban planning to map informal settlements, identify infrastructure gaps, and enhance service delivery (Melnyk et al., 2020).

Although such applications reveal how space-based technologies can be used in development, much of the literature focuses on technical potential and policy impacts. There is very little interaction with the social aspects of these technologies, especially regarding who benefits from them and how they affect marginalized groups.

Artificial Intelligence, Data Governance, and the Digital Divide

The inclusion of the concept of artificial intelligence (AI) in development practices further altered the face of poverty reduction. Predictive analytics, welfare targeting, and resource allocation are increasingly automated by AI-driven systems, providing new opportunities to achieve efficiency and scale (Jacobs et al., 2020). But there are also serious doubts about the social consequences of these technologies, as scholars have suggested.

According to Kitchin (2014), the data revolution has resulted in novel forms of governance that depend on data-driven decision-making processes, and this situation is often not accompanied by appropriate consideration of accountability and transparency issues. Likewise, Couldry and Mejias (2019) introduce the notion of data colonialism to describe how the extraction and control of data might reproduce power dynamics, especially between the developed and developing parts of the world.

The digital divide remains a major obstacle to the fair application of AI and geospatial technologies. Digital infrastructure, technical expertise, and data literacy are not equally distributed, as they tend to bypass the very groups of people that such technologies are supposed to support (Van Dijk, 2020). Communities residing in rural and marginalized areas often lack access to reliable internet or digital resources in the Indian context, thereby restricting their involvement in technology-focused development programs.

Gaps in Existing Literature

The existing literature on poverty, spatial inequality, and technological innovation continues to have numerous gaps despite the increase in publications. To begin with, sociological theories of inequality and methods of technological development have not been integrated in an interdisciplinary manner. Concrete research addresses the power of space-based and AI-driven systems, but does not always address poverty in its structural aspects.

Second, the literature is largely top-down, oriented to the implementation of policies and the institutional system rather than the lives of the working poor. This restricts our knowledge of the perception, access and use of technological interventions by the marginalized groups.

Third, insufficient focus is given to participation and knowledge production issues. The majority of geospatial and AI-based systems rely on data gathered and analyzed by professionals, with few local communities involved. This raises the issue of representation, as the knowledge and views of marginalized populations are often sidelined in decision-making.

Lastly, a normalization framework is necessary to incorporate aspects of equity, inclusion, and justice into the development and use of technological interventions. As much as the idea of spatial justice is a good starting point, its exploration in relation to space-based technologies and the alleviation of poverty is not well developed.

Positioning of the Present Study

To address these gaps, this chapter takes an interdisciplinary perspective, integrating sociological theory and technological analysis. Placing space-based technologies in the context of spatial justice, it attempts to relinquish technocratic discourses and preempt the lives of the working poor. The chapter fills the current literature with a proposal for a sociological framework that focuses on accessibility, participation, equity, and the integration of local knowledge systems.

THEORETICAL FRAMEWORK

Spatial Justice and the Geography of Inequality

Spatial justice is a notion that offers a critical approach to determining how inequalities are generated, structured, and reproduced through space. Going beyond conventional ideas of justice, with their social or economic orientation, the concept of spatial justice prioritizes the geographical character of inequality and the influence of spatial forms on access to resources, opportunities, and power (Barnett, 2018). It questions the notion that space is a neutral frame of reference for social processes. It suggests that the relations of politics, the economy, and social interactions actively construct space.

The theorization of spatial justice dates back to the studies by (Dikeç, 2001), who discussed space as a social construction subject to power and production relationships. The concept of the production of space by Lefebvre emphasizes that the spatial organization of cities, in the form of urban planning, land use, and infrastructure development, is integrated into larger systems of capitalism and authority. Spatial structures can favor some groups over others, leading to uneven development and marginalization. Developing this, (Harvey, 2010) says that the spatial inequalities are not accidental but rather produced systematically in the course of capital accumulation, in which investments are made in areas that are profitable, and the less productive areas are left alone.

The perspective of (Soja, 2010) goes a bit further and expressly connects the spatial structure with the issues of method and justice. Soja defines spatial justice as the equitable distribution of socially prized resources across space and equitable access to opportunities regardless of geographical location. It highlights how marginalized people are usually restricted to areas with inadequate infrastructure, services, and economic potential. In this, spatial injustice is a cause and effect of other social injustices, strengthening poverty and exclusion cycles.

Spatial justice is an important prism through which to examine how poverty is geographically constituted among the working poor. The poor are highly clustered in structurally disadvantaged spaces, such as distant rural areas, informal urban slums, and ecologically sensitive zones. Such spaces are usually characterized by poor infrastructure, poor connectivity, and limited access to education, health, and job opportunities. For example, communities in rural areas can experience difficulties due to geographic seclusion and limited market access. In contrast, residents of slums in cities face overcrowding, tenuous housing, and a status as non-objects of formal planning. It is not only that such spatial conditions reproduce existing inequalities, but it is also the factor that leads to their reproduction.

Spatial justice is especially important nowadays, when space-based technologies and data-driven governance are a part of a new era. There are also emerging opportunities to use satellite imagery, GIS, and remote sensing to map and address spatial inequalities, enabling targeted interventions and resource distribution. Nevertheless, the use of these technologies does not necessarily yield fairer results. As (Kitchin, 2014) notes, the use of data focuses on efficiency and optimization, which may neglect the political aspects of inequality. Spatial data can reveal patterns of deprivation, but it does not necessarily address the power relations underlying those patterns.

Furthermore, the processes of spatial data production and control are also circumscribed by systems of power. State agencies, corporations, or technical experts usually make decisions about which data to gather, interpret, and use, with little or no consultation with isolated communities (Couldry & Mejias, 2019). This raises important questions about representation and inclusion, as the lived experiences and local knowledge of the working poor are often excluded from formal data systems. Consequently, space-based technologies unwillingly reproduce spatial injustices through strengthening top-down approaches to development that do not address the realities of communities.

Meanwhile, spatial justice offers new prospects for a more inclusive, participatory approach to development. The merging of participatory GIS and community-based mapping projects demonstrates the potential of spatial technologies to empower marginalized populations by enabling them to map their own spaces and articulate their needs. These strategies are consistent with the principles of spatial justice, as they help provide equal access to information, increase people's involvement in decision-making, and disrupt hegemonic discourses about space and development.

In this sense, spatial justice plays a crucial role in bridging the gap between sociological theory and technological practice. It highlights the importance of leaving behind technocratic conceptualizations of poverty reduction and of adopting approaches that could facilitate the use of space-based interventions designed and implemented in ways that would facilitate equity, inclusion, and social change. With future predictions of the spatial dimensions of inequality, this framework provides a basis for a critical assessment of the role of emerging technologies in addressing the conditions of the working poor.

Capability Approach and Technological Inclusion

The capability approach, developed by Sen (1999), provides a revolutionary way of looking at poverty in terms of a human being rather than based on income levels. Instead of just understanding poverty by economic deprivation, Sen Construct's poverty as a deprivation of capabilities, the substantive freedoms that people have to live lives that they have a reason to cherish. This viewpoint shifts the resources-and-production-oriented ideology to prospects and results, as they are actually in a position to do and become. From this, one can see that poverty is not simply a deprivation of income but a multidimensional state shaped by limitations on access, choice, and agency.

The ability approach holds that resources and capabilities are distinct. Though resources, i.e., income, infrastructure, or technology, matter, they do not necessarily make a difference in well-being. Transforming resources into capabilities is determined by various personal, social, and environmental factors, as well as education, social norms, institutional arrangements, and geographic conditions. For the working poor, such conversion factors are usually highly unequal, thereby reducing their opportunities to benefit from development interventions when such resources are available.

With space-based technologies and artificial intelligence, the capability approach will offer a critical view through which technological innovations can be evaluated to determine whether they truly improve the lives of disadvantaged populations. Even though technologies such as GIS, remote sensing, and AI-based analytics can produce useful information and improve the efficiency of service delivery, their impact on poverty reduction will depend on whether they increase the true freedoms of the working poor. An example is the use of satellite-based agricultural advisories to enhance crop productivity, which depends on Without these enabling conditions, technological interventions risk remaining inaccessible or irrelevant to those most in need.

This is a major drawback of technocratic development strategies, which often assume that offering technological solutions will automatically lead to good outcomes. Capability-wise, these assumptions have a problem insofar as they do not take into account the social and structural limitations that created opportunities to access and use technology. Sen (1999) points out that development should not be considered solely in terms of technological progress but also in terms of its effects on human freedoms. The implementation of space-based technologies, in this respect, should be evaluated in terms of their capacity to enhance capabilities, including access to information, participation in decision-making, and control over resources.

Another aspect highlighted by the capability approach is agency, which is the power of individuals to accomplish objectives that they hold in esteem. The working poor have become easy targets and recipients of aid in monology-based development projects. Still, they are not directly involved in planning. This agency deficiency undermines the success of these and may lead to a gap in logical solutions and local solutions. To illustrate, geospatial mapping of rural regions can also reveal resource disparities. Unless the community participates, interventions based on the findings may not account for context-specific issues and cultural constraints.

Moreover, the introduction of artificial intelligence into governance systems also creates other issues, such as the deprivation of capability. Most AI-based decision-making systems are also based on large datasets, which may not reflect the experiences of marginalized groups, leading to biases in resource distribution and policy implementation (O'Neil & Ucbasaran, 2016). Such systems may have the unintended impact of strengthening existing inequalities, as they are built on the principle of efficiency over equity and thus crippling the potential of people who already have a disadvantage. These challenges are further exacerbated by the lack of transparency and accountability in algorithmic systems, raising questions about fairness and justice in technology governance.

Meanwhile, the capability approach provides a normative basis for designing more inclusive and equitable technological interventions. It underscores the need to establish enabling conditions for people to use technological resources successfully. These encompass investment in digital literacy, infrastructure, and institutional support, as well as the introduction of local knowledge and community involvement in the development processes. This practice of emphasizing capabilities over resources promotes a transition from top-down, technology-based models to more humanistic, participatory models.

The applicability of the capability approach is especially strong in the Indian context, where there are huge inequalities in access to education, technology, and infrastructure. Rural and marginalized groups frequently have several levels of deprivation that restrict their capacity to enjoy technological advances. To close these gaps and ensure that technological advances translate into substantial improvements in well-being, one may integrate space-based technologies with programs that strengthen capabilities (through education, skills development, and community participation).

To sum up, the capability approach provides a critical paradigm on how space-based technologies and AI can be used to reduce poverty. It emphasizes that technological advancement alone is not enough to address the complex realities of the working poor. Rather, development interventions should be structured to increase human capabilities, facilitate agency, and address structural inequalities. By incorporating this approach, the current research transcends technocratic discourse and underscores the necessity of a more inclusive, socially grounded mode of technological development.

Social Exclusion, Digital Divide, and Data Inequality

Social exclusion is a concept that offers an important perspective on how some groups are systematically denied access to resources, opportunities, and involvement in societal processes. In contrast to the classical conceptualizations of poverty that emphasize economic deprivation as the main characteristic, social exclusion stresses relational and structural concepts. It concerns the exclusion of individuals and groups from systems across social, political, and financial dimensions (Fang et al., 2019). This view is especially applicable when examining the experiences of the working poor who usually encounter several and compounding types of exclusion founded on class, caste, gender, ethnicity, and geographic location.

In modern development, social exclusion as a process is increasingly mediated by digital technologies. The rapid adoption of information and communication technologies (ICTs), artificial intelligence (AI), and space-based systems has transformed governance and service delivery, creating new channels for efficiency and innovation. Nevertheless, progress has also brought new forms of inequality, often referred to as the digital divide. Digital divide is not only an inequality in access to technology but also a difference in access to, use of, and benefits from digital systems.

For the working poor, the digital divide operates at multiple levels. First, it is the inherent disparity in access to digital infrastructure, such as reliable internet connections, electricity, and digital devices. This has been more pronounced in rural and remote regions, where a lack of infrastructure restricts the coverage of technology-based interventions. Second, digital literacy and skills disparities also exist and influence individuals' prospects for accessing digital platforms and using technological resources productively. Third, and arguably most importantly, representations and participation in data systems are unequal, with the voices and experiences of marginalized groups frequently absent or distorted.

Such aspects of the digital divide are important to the application of space-based technologies and AI in reducing poverty. Although GIS, remote sensing, and satellite data may help provide specific knowledge of the spatial pattern of deprivation, they tend to rely on top-down data collection and analysis methods that do not involve local communities. Consequently, the knowledge generated by such systems might not be grounded in the lived realities of the working poor, leading to interventions that are not suitable to local needs and priorities (Kitchin, 2021). This disconnect underscores the limitations of purely data-driven approaches to development, which may overlook cultural contexts within which poverty is experienced.

The notion of data inequality further deepens this critique. Couldry and Mejias (2019) suggest that the modern data economy is built on data colonialism, in which data are harvested from people in a substantial way, with no significant consent or value. Within this structure, marginalized communities are reduced to sources of information rather than knowledge producers. Data control and ownership are left to state institutions, corporations, and technical specialists, which reinforces the status quo of power imbalance. In the case of the working poor, this may lead to a paradoxical situation in which they become more apparent in data sets but less so in decision-making.

These considerations are further enhanced by artificial intelligence, as algorithmic systems are often trained on datasets that are already biased and unequal. O'Neil (2016) argues that algorithmic decision-making contributes to discrimination by introducing past biases into predictive models. This can manifest as an unequal distribution of resources for poverty alleviation, disqualification from welfare programs, or categorization as vulnerable. These problems are also complicated by the fact that AI systems lack transparency and do not allow one to dispute or fix biased results.

Social exclusion and the digital divide in the Indian context aim at a point of intersection with already existing structural inequalities such as caste hierarchies, gender inequalities and regional imbalances. The disadvantaged communities, especially in rural and marginalized community locations, are usually compounded, and this restrains their access to both physical and digital structures. Although efforts to fill these gaps, such as the Digital India initiative, have reduced them, significant gaps remain in access, usage, and outcomes. These issues underscore the importance of a more subtle approach to the nature of interactions between technological interventions and the social structures already in place.

Simultaneously, the literature also identifies the possibility of using participatory and community-based approaches to address these inequalities. Activation of marginalized populations. They may undermine top-down models of data governance and align with the broader objectives of inclusiveness and social justice.

Finally, social exclusion, the digital divide, and data inequality are concepts that offer an essential understanding of the limitations and challenges of technology-driven poverty alleviation. They emphasize that access to technology is not enough to guarantee fair outcomes and that concerns about power, representation, and participation should be paramount in any study of technological interventions. The chapter's addition of these views highlights the necessity of a more inclusive, socially grounded approach to the application of space-based technologies and AI to address the plight of the working poor.

Working Poor and Spatial Marginalization

This is because the situation of the working poor can never be fully understood without placing it within the context of inequality and spatial organization. As the working poor are already involved in economic activity, their livelihoods tend to be in the structurally disadvantaged spaces that restrict access to resources, services, and opportunities. Such spatial limitations are not accidental but are fundamentally connected with broader processes of socio-economic life that influence development and exclusionary trends. Therefore, the problem of poverty among the working poor must be considered not only in terms of income or occupation but also in terms of spatial marginalization.

The concept of spatial marginalization refers to mechanisms that push some groups of people into geographically disadvantaged places with poor infrastructure, low connectivity, and reduced access to institutional services (Soja, 2010). These areas, spanning remote rural areas to informal urban areas, are usually not included in mainstream development planning, and thus, there will always be differences in the standard of living. To the working poor, these spatial disadvantages are in the form of decreased access to education, healthcare, markets and employment opportunities, thus perpetuating poverty cycles.

In rural settings, spatial marginalization is directly linked to geographic isolation and a shortage of infrastructure. Most rural people, especially in developing nations such as India, reside in areas with strong transportation systems but poor access to markets and government services. This alienation restricts economic mobility and reduces laborers' wages, even when people are actively engaged in agricultural or non-agricultural activities. Inability to access timely information, credit facilities, and market connections also increases the vulnerability of small and marginal farmers, making it hard for them to escape poverty. There are also environmental influences, such as climate change and resource degradation, which disproportionately impact the rural population, increasing their spatial and economic precarity.

There is also a great manifestation of spatial inequality in urban spaces, which are commonly viewed as places of opportunity. Cities in the Global South have been growing fast, causing the spread of informal settlements and slums, in which a significant percentage of the working poor dwell. The areas are normally overcrowded, insecurely housed, poorly sanitized, and lack access to basic amenities. Although they are located close to economic centers, people in informal settlements are usually marginalized in planning processes. They are not recognized by law, and therefore have minimal access to communal resources and social protection networks. This urban poverty, emphasizing spatial exclusion, underscores the irony of the city as a location of poverty, because geographical location may not always correlate with a higher quality of life.

One of the most significant aspects of spatial marginalization in the Indian context is that of marginalized and indigenous communities. Such populations are commonly found in geographically remote and ecologically vulnerable areas, such as forested and hilly areas, abundant in natural resources but impoverished in infrastructure. Traditionally, these populations have been excluded from mainstream economic and political activities, resulting in limited access to education, health care, and employment opportunities (Xaxa, 2011). Meanwhile, since they rely on natural resources, they are susceptible to environmental shifts and displacement by development projects. The spatial segregation of marginalized groups is therefore an intricate combination of geographical, economic, and socio-political factors that continue to marginalize them.

Spatial marginalization is deeply interconnected with social exclusion processes in a sociological view. Marginalized spaces are likely to be associated with marginalized populations wherein caste, class, gender, and ethnicity intersect to create compounded forms of disadvantage. As an example, the groups of lower castes and migrants are overrepresented in informal urban settlements, whereas the populations of the marginalized communities are overrepresented in remote regions. These spatial distributions are not arbitrary but the result of historical forces of discrimination, land alienation and unequal development (Desai & Dubey, 2012). As a result, spatial marginalization both reflects and reinforces existing social hierarchies.

Some of the issues of spatial marginalization could be addressed through the growing application of space-based technologies and artificial intelligence to development planning, thereby enhancing the identification and targeting of vulnerable populations. For example, geospatial mapping can be used to identify underserved regions, track infrastructure gaps, and support more efficient resource allocation. Nevertheless, as mentioned in other sections, such technologies can also be dangerous if they are not developed and adopted inclusively, as they can reproduce existing inequalities. Unless the underlying concerns of access, participation, and representation are addressed, technological interventions might not reach the most marginalised groups or misjudge their needs.

In this regard, it is important to note that the condition of the working poor should be approached in a multidimensional manner that synthesizes spatial analysis with sociological knowledge. It goes beyond making infrastructure and services more accessible; it is also about questioning the structural processes that generate spatial inequalities. This chapter situates the working poor within the context of spatial marginalization to highlight the importance of development strategies that are geographically sensitive and socially inclusive.

Space-Based Technologies in Poverty Alleviation: Potentials and Limitations

This has changed how poverty is measured, understood, and addressed, due to the growing use of space-based technologies in development planning. Geographic Information Systems (GIS), remote sensing, satellite imagery, and, more recently, artificial intelligence (AI)-based geospatial analytics technologies have enabled governments and international agencies to produce fine-grained spatial data, observe environmental and socio-economic processes, and formulate targeted interventions. These instruments are also usually marketed as creative initiatives that can boost productivity, accuracy, and the scale of poverty relief programs (Monico, 2004). Nevertheless, their potential is immense, and they can only have an effect depending on the social and political environments in which they are used.

APPLICATIONS IN AGRICULTURE AND LIVELIHOODS

Among the most notable uses of space technologies has been in the agricultural sector, in which a significant share of the working poor, especially in developing nations such as India, have found work. Remote sensing using satellites is commonly employed to assess crop health, measure soil moisture, forecast weather, and estimate yields. With these technologies, farmers and policymakers can make sound choices about irrigation, fertilization, and crop selection, enhancing productivity and resilience.

Moreover, geospatial tools also enable precision agriculture and hence more efficient utilization of resources, e.g. water and fertilizers. For smallholder farmers, the risks posed by climate change and market fluctuations can be greatly reduced by accessing timely, accurate information. Such technologies have become more popular among governments and development agencies as a complement to agricultural extension services and have been used to enhance food security.

The advantages of such technologies, however, are not evenly spread. The small and marginal farmers, who often comprise a large portion of the working poor, are hindered from accessing and using such information due to low digital literacy, infrastructural barriers, and financial constraints. Consequently, although the use of space-based technologies can positively impact agricultural productivity, they can also exacerbate existing inequalities by limiting access for farmers with fewer resources.

Disaster Management and Climate Resilience

Space technologies are very important for disaster management and reducing climate risks, particularly for the working poor, who are more often affected by environmental risks than others. A real-time track of natural calamities such as floods, droughts, cyclones, and forest fires can be monitored using remote sensing and satellite imagery. Using these tools, it is possible to implement early warning systems, risk mapping, and damage assessment, thereby making disaster preparedness and response more effective.

For example, satellite-based flood mapping would enable the identification of vulnerable areas and support timely evacuation and resource distribution. On the same note, drought surveillance systems can guide water management policies and support agricultural planning. These technologies help build resilience in areas susceptible to climate change by providing essential information that can help mitigate the effects of environmental shocks.

Although these benefits are enjoyed, the success of such interventions depends on how deeply the information penetrates and is internalized by affected communities. In most instances, early warning systems do not lead to proper action due to communication disconnects, a lack of trust in institutions, or a lack of socio-cultural advantages. Besides, disaster response initiatives tend to be more technologically efficient than community-focused, which limits their long-term usefulness.

Urban Planning and Informal Settlements

Space-based technologies have also been utilized more extensively in city settings to map and govern informal settlements, where a large proportion of the working poor live. GIS and satellite images allow one to establish the location of slums, evaluate infrastructure deficits, and design interventions such as these. These instruments provided useful information for an urban environment, for policy, and for developing specific development to address living conditions.

Take the example of geospatial mapping, which can be used to identify areas without access to basic services and target them for development initiatives. It is also capable of supporting land-use planning and the integration of informal settlements into formal urban systems. In this regard, it is possible to argue that space-based technologies should enable more inclusive, data-driven urban development.

Nonetheless, surveillance, displacement, and exclusion also become issues when such technologies are used. The informal settlements being mapped may at times result in their eviction or redevelopment without proper consideration of their rights and livelihoods. Moreover, data-based classification of spaces can simplify and erase the complexity of informal communities, which are diverse and evolve. These issues demonstrate the necessity of ethical, participatory planning for cities.

Welfare Targeting and Governance

The other vital use of space-based and AI-driven technologies is in welfare targeting and governance. By combining geospatial data with socio-economic indicators, governments can pinpoint vulnerable groups and allocate resources more quickly. For example, poverty areas can be mapped using satellite data, infrastructure development can be traced, and welfare programs can be tracked.

Nonetheless, overreliance on data-driven systems also poses serious risks. The process of algorithmic decision-making can be biased by the lack of complete or representative data, thereby causing exclusion errors or resource misallocation (O'Neil & Ucbasaran, 2016). Moreover, transparency in AI systems is lacking, so accountability and fairness are questioned. These problems can be quite severe for the working poor, where misuse of information or algorithms could lead to a refusal of necessary services.

Critical Limitations and Sociological Concerns

Concentration of control over technological systems is another major problem. Space-based technologies are usually developed and managed by state agencies, the corporate sector, and other technical professionals, with little engagement from oppressed communities. This creates power imbalances in decision-making and casts doubt on whose interests are being served (Couldry & Mejias, 2023).

Moreover, overdependence on quantitative data may push out qualitative information, especially the native and local knowledge systems that are important for interpreting particular situations. To illustrate, marginalized communities tend to have comprehensive ecological information that cannot be reflected by the satellite images but can be used to ensure the sustainable use of resources. Undergoing such knowledge may lead to ineffective, even damaging interventions.

Lastly, the digital divide remains a major obstacle to the fair application of space-based technologies. As noted in the above discussion, the inequalities in access to digital infrastructure, skills, and resources prevent the working poor from enjoying the benefits of technological advancements. The implementation of high-tech solutions can only aggravate the poverty problem without addressing the underlying inequalities.

Toward Inclusive Technological Interventions

To realize the untapped potential of space-based technologies for poverty alleviation, a more inclusive, socially grounded approach should be embraced. This will incorporate technological advances with the concepts of spatial justice, capability building, and participatory governance. This is important for ensuring fair access to data, community participation in decision-making, and the integration of local knowledge systems.

Also, the use of AI and geospatial technologies must be more transparent and accountable. Ethical protocols, regulatory frameworks, and interdisciplinary teamwork can be used to address some of the issues associated with data-driven governance. It can be suggested that one can utilise the transformative potential of technological tools and reduce their risks by aligning technological interventions with broader objectives of social justice and inclusion.

CASE-BASED INSIGHTS (INDIA FOCUS)

The Indian experience with space-based technologies provides an empirically rich basis for analyzing their contribution to poverty alleviation among the working poor. In recent decades, India has become increasingly involved in the consideration of geospatial technologies for governance and development planning, with initiatives led by the Indian

Space Research Organization (ISRO) and other related organizations. These programs cover various industries, including agriculture, rural development, disaster management, and urban planning. Although they reveal the spatial potential of tools grounded in space, they also show the limitations that are crucial to consider through the prism of spatial justice and social inclusion.

ISRO and Geospatial Applications in Development

The space program in India has been instrumental in the exploitation of satellite technologies for socio-economic development. The satellites of ISRO provide high-resolution data, which is utilised for land-use mapping, tracking natural resources, and enabling developmental planning. The development of geospatial databases has been made possible through programs such as the National Remote Sensing Centre (NRSC) initiatives, which are used for decision-making across different sectors (Viso, 2023).

Nevertheless, these technologies enhance the capacity of the states, but the working poor do not receive such advantages equally. Such platforms are also digital and demand technical expertise that is unavailable in marginalized rural regions. Thus, geospatial data is still used by institutional actors and does not allow direct interaction with local communities.

GIS in Rural Planning and Development

The use of Geographic Information Systems (GIS) has been growing within the rural planning program to increase the efficiency and effectiveness of the development program. GIS-based planning in India has been applied to watershed management, rural infrastructure, and land resource mapping schemes. For example, geospatial tools have been employed to track the creation of assets and to ensure accountability for implementation under the Mahatma Gandhi National Rural Employment Guarantee Act (MGNREGA).

GIS allows the identification of resource-deficient zones, the evaluation of land-use patterns, and the design of interventions that include water-curtailing buildings and roads. Such applications can resolve spatial inequalities by targeting resources to undervalued areas. Ideally, these data-driven solutions can improve targeting and reduce regional inequalities.

Disaster Mapping and Risk Reduction

India's vulnerability to natural disasters has led to the widespread application of space-based technologies for disaster management. Flood mapping, cyclone tracking, drought evaluation, and damage assessment are among the applications of satellite imagery and remote sensing. Geospatial data is used by institutions such as ISRO and the National Disaster Management Authority (NDMA) to support early warning systems and coordinate response efforts.

In cases such as recurring floods in other states, such as Assam and Bihar, satellite flood mapping has been employed to detect affected areas, assess damage, and provide relief. Equally, remote sensing data-based drought monitoring systems assist in identifying and mapping water-stressed areas and in developing mitigation strategies. As shown by these applications, space-based technologies can improve preparedness and minimize vulnerability among vulnerable populations.

Nevertheless, such data is not evenly translated into effective outcomes for the working poor. Even early warning systems are not always effective, as communication fails, language is misinterpreted, and access to information channels reaches a remote population that is often the most marginalized. Also, disaster response measures are often focused on short-term relief and do not consider long-term rehabilitation, thereby disregarding the structural factors that predispose certain groups to disasters. Consequently, even though technological devices enhance institutions' capacity to respond, their effect on alleviating vulnerability among the working poor is mediated by the overall socio-economic context.

Indigenous Communities and Local Knowledge Systems

Another vital aspect that is not usually considered in technology-based development is the indigenous and local knowledge system. Marginalized communities in India are found in geographically isolated areas, such as forests, hills, and ecologically sensitive areas. These communities are well-versed in local ecosystems, farming techniques, and how resources are managed over generations.

There has been extensive use of space-based technologies, especially for forest monitoring and land-use mapping, in these regions. Although these interventions are significant for conserving the environment and planning, they are usually centralized and exclude indigenous perspectives. For example, satellite forest mapping can categories land by vegetation and fail to identify the cultural and livelihood value of such spaces to marginalized populations.

This disconnection may result in disagreements over land use and resource rights, especially when technological data is utilized to support conservation or development projects without sufficient consultation with the people. Besides, the marginalization of indigenous knowledge by formal data systems constitutes epistemic marginalization, that is, the depreciation or disregard of local forms of knowing.

Meanwhile, the potential for combining geospatial technologies with participatory strategies is increasingly recognized. Community-based mapping programs, in which local communities participate in describing and interpreting spatial information, provide a more comprehensive approach to development. These strategies not only lead to accurate data but also empower communities by giving them a voice in decision-making.

Synthesis: Bridging Technology and Social Reality

The Indian case shows that space-based technologies are effective strategies for addressing spatial inequalities, but social, institutional, and cultural contexts strongly influence the success of technological interventions. Across industries, there is a general trend: technological interventions are highly effective when combined with local knowledge, community involvement, and inclusive governance systems.

On the other hand, top-down strategies that focus on technical efficiency rather than social inclusion usually do not address the working poor. The dependence on centralized data systems, the lack of community participation, and inadequate focus on the local context may lead to the production of interventions that are misaligned with ground realities.

These observations necessitate a more comprehensive approach that integrates technological innovation and sociological knowledge. With the introduction of the concepts of spatial justice, the power of capability improvement, and participatory governance, it can be designed to create an intervention that is efficient, yet fair and inclusive.

SOCIOLOGICAL FRAMEWORK FOR SPACE-BASED POVERTY ALLEVIATION

The above discussion has demonstrated an important contradiction in modern development practice: space-based technologies and the application of artificial intelligence provide potent means to mitigate poverty, but their applicability is heavily dependent on the social environments in which they are deployed. Current methods tend to focus on technical efficiency, data optimization, and large-scale governance of the working poor, without considering structural inequalities and the lived realities of the working poor. To address these constraints, this chapter proposes a sociological approach to space-based poverty reduction that combines technological innovation with ideas of spatial justice, capability building, and social inclusion.

The model is anchored in four interconnected dimensions: accessibility, participation, equity, and knowledge integration. These dimensions combine to offer an integrated approach to learning and applying technology-driven interventions that are effective and socially just.

Accessibility: Bridging Technological and Spatial Divides

Accessibility is the first dimension of the framework, underscoring the importance of ensuring that space-based technologies are accessible and accessible to the working poor. As mentioned above, digital inequalities, connectivity, and literacy are major barriers to enabling marginalized groups to enjoy the benefits of technological advances (Van Laar et al., 2020). Unless these barriers are addressed, the introduction of sophisticated technologies will only deepen existing disparities.

From a sociological point of view, accessibility goes beyond the physical accessibility to include the ability to have a meaningful interaction with the technology. This includes investment in digital literacy, localized information systems, and easy-to-use interfaces that support the various populations. The design of available technologies should be responsive to local conditions and needs in rural and marginalized communities, where linguistic and educational barriers are common.

Moreover, accessibility is intimately connected to spatial justice, as it concerns the uneven distribution of technological resources across geographic space. It is necessary to ensure that the working poor have equal opportunity to participate in development processes and to enjoy technological progress by providing equitable access to space-based tools and data.

Participation: Democratizing Data and Decision-Making

The second dimension of the framework focuses on participation, underscoring the need to engage marginalized communities in data production, interpretation, and decision-making processes. Conventional methods of space-based technologies are usually typified by top-down governance, in which data are gathered and processed by experts with little input from local populations.

The challenge to this model is participation, which is proposed as a more inclusive and democratic mode of knowledge production. Examples of ways local communities can actively acquire and use spatial data include participatory GIS and community mapping projects (Elwood, 2006). These measures not only make data more relevant and accurate but also empower communities by giving them a voice in development planning.

Participation is critical among the working poor to ensure that technological interventions are based on local needs and priorities. It helps communities express their experiences of marginalization, identify issues unique to their context, and develop solutions together. Participatory practices enhance sustainable, more effective development by fostering agency and ownership.

Equity: Ensuring Fair Distribution of Benefits

The normative core of the suggested framework consists of equity, which aims to ensure the just allocation of resources, opportunities, and the benefits of technology interventions. Although space-based technologies can enhance efficiency and targeting, they do not automatically yield fair results. Such technologies will likely further entrench the unequal benefits of already advantaged groups without conscious attempts to address structural disparities.

In terms of spatial justice, equity entails redressing imbalances in the allocation of infrastructure, services, and opportunities across regions. Practically, this should take the form of prioritizing the implementation of interventions in underserved areas, ensuring that marginalized populations are integrated into development programs, and addressing systemic barriers to accessing resources.

Equity also applies to data governance issues such as transparency, accountability, and fairness in AI and geospatial use. The algorithms should have mechanisms to reduce bias and avoid omissions, especially when the outcomes are welfare-targeting, because mistakes could be very costly in the real lives of vulnerable populations. To achieve fair results, technical protection alone is not enough; ethical and institutional systems with specific priorities for social justice are also needed.

Knowledge Integration: Bridging Scientific and Indigenous Systems

The fourth dimension of the framework focuses on how various systems of knowledge are incorporated, especially indigenous and local knowledge, into technological interventions. As mentioned earlier in this paper, space-based technologies tend to use centralized data systems that can fail to capture the subtlety and contextual awareness of local communities.

The native knowledge regimes, particularly those of marginalized communities, provide insightful information on ecological processes, resource management, and sustainability. This knowledge can be integrated with geospatial information to promote the accuracy, relevance, and sustainability of development interventions. For example, an integrated approach combining satellite-based environmental monitoring with local observations can provide a clearer understanding of climate variability and resource consumption.

Knowledge integration is also concerned with the issues of epistemic justice in that knowledge of various kinds is treated and esteemed. It also questions the prevailing technocratic leadership and encourages a more pluralist and inclusive approach to development. This dimension helps bring together scientific and local knowledge systems, thereby developing solutions that are more context-sensitive and culturally appropriate.

Synthesis of the Framework

The combination of the four dimensions accessibility, participation, equity, and knowledge integration gives a holistic sociological approach to the analysis and application of space-based poverty alleviation plans. All these dimensions cannot exist in isolation, but are closely interrelated, supporting each other in a quest towards inclusive and fair development.

Indicatively, increasing accessibility, but not participation, can lead to a lack of technology utilization, whereas participation, but not equity, can be ineffective in addressing structural inequalities. Likewise, knowledge integration improves participation and equity by representing different perspectives in the decision-making process. The power of the framework lies in its inclusion of these dimensions in a consistent way that supports technological innovation and social justice.

Implications of the Framework

The suggested framework presents significant consequences of research and practice. It demands a shift from strictly technocratic development models to models based on sociological interpretation and morality. To policymakers, this translates into the development of interventions that place a high priority on inclusivity, transparency, and community involvement. To researchers, it points to the necessity of interdisciplinary studies that would integrate the knowledge of sociology, geography, and technology research.

Finally, this framework highlights that space-based technologies are only effective at alleviating poverty when they operate under both technical and social conditions. It uses the experiences and needs of the working poor to create an avenue for exploiting technological innovation that fosters equity, inclusion, and sustainable development.

POLICY IMPLICATIONS: TOWARD INCLUSIVE TECHNOLOGICAL GOVERNANCE

The discussion in this chapter shows that the efficiency of space-based technologies and AI in reducing poverty depends on both technical efficiency and inclusive and equitable governance. Policy interventions should overcome structural inequalities, access barriers, and participation barriers to ensure that these technologies can serve the working poor.

To begin with, it is necessary to expand digital inclusive infrastructure. The persistent disconnects in internet access, power, and digital devices continue to isolate rural and marginalized populations (Van Laar et al., 2020). Policymakers should focus on investments in underprivileged areas and increase digital literacy, as this would enable communities to interact effectively with geospatial and artificial intelligence-based systems.

Second, participatory governance has to be institutionalized. Development interventions are frequently top-down and do not take into account local realities. Participatory GIS and community mapping can be used to democratize data production and ensure that the working poor are not excluded from decision-making (Elwood, 2006).

Third, ethical and responsible data control should be enhanced. The growing use of AI is associated with bias, transparency, and exclusion. Regulations must be fair concerning algorithmic decision-making, safeguard privacy, and create accountability mechanisms (O'Neil & Ucbasaran, 2016).

Fourth, the policies should focus on equity in resource distribution. Although space-based technologies allow reaching the target accurately, they should be used to eliminate, rather than support, spatial inequalities. This involves targeted targeting of marginalized areas, such as rural communities and informal towns.

Finally, consideration of local and indigenous knowledge systems is important for planning context-sensitive interventions. The realization of marginalized and rural communities' knowledge can lead to the relevance and sustainability of development programs.

In short, there should be a move towards a justice-based development practice that balances technological innovation with equity, inclusion, and participation. That is when space-based technologies can have a significant impact on poverty alleviation of the working poor.

CONCLUSION

The chapter has discussed the application of space-based technologies and artificial intelligence to alleviate poverty from a sociological perspective, with a special focus on the plight of the working poor. As technological innovations (including GIS, remote sensing, and AI-based analytics) have greatly improved the ability of states and institutions to map and trace poverty, as well as mitigate it, the effects of technology are not evenly distributed. The existing social and spatial inequities often constrain them.

The chapter has shown that poverty is not just an economic status but a multidimensional phenomenon, as it is entrenched in geographic, social, and institutional structures, as evidenced by the concepts of spatial justice, the capability approach, and social exclusion. The working poor tend to be situated in marginalized areas, whether rural, informal, or otherwise, where services, infrastructure, and opportunities are limited. In this regard, technological interventions, when applied without consideration of such structural realities, are likely to reproduce rather than relieve existing disparities.

This tension is also demonstrated in the Indian context analysis. Although efforts by institutions like ISRO and the application of geospatial technologies in agriculture, disaster management, and urban planning have demonstrated significant potential, the processes of access, participation, and representation play an intervening role in effectiveness. The techno-centric system of development can also be seen in the fact that marginalized population groups are not represented in data systems and decision-making processes.

In response, this chapter has suggested a sociological framework to be applied across four major dimensions: accessibility, participation, equity, and integration of knowledge. This framework emphasizes that technology should be integrated into socially inclusive, context-oriented development strategies. It demands closing digital gaps, data democratization, equitable distribution of benefits and incorporation of indigenous knowledge regimes into technological interventions.

Finally, the chapter states that space-based technologies can serve as a useful means of poverty reduction only when justice and inclusion are prioritized. Instead of conceiving of technology as a neutral or purely technical solution, it is crucial to recognize that it has social and political aspects. There should be a paradigm shift to a more interdisciplinary and humanistic approach in which the experiences and abilities of the working poor are at the center stage, rather than peripheral, to achieving sustainable and equitable development outcomes.

REFERENCES

Alkire, S., & Santos, M. E. (2013). A multidimensional approach: Poverty measurement & beyond. *Social Indicators Research, 112*(2), 239–257. DOI: 10.1007/s11205-013-0257-3

Barnett, C. (2018). Geography and the priority of injustice. *Annals of the Association of American Geographers, 108*(2), 317–326.

Citaristi, I. (2022). United nations office for outer space affairs—unoosa. In *The Europa Directory of International Organizations 2022* (pp. 247–248). Routledge. DOI: 10.4324/9781003292548-53

Constantin, D.-L. (2021). Addressing spatial justice at lower territorial levels. some insights from the central and east European countries' perspective. *Regional Science Inquiry, 13*(2), 315–326.

Couldry, N., & Mejias, U. A. (2019). Data colonialism: Rethinking big data's relation to the contemporary subject. *Television & New Media, 20*(4), 336–349. DOI: 10.1177/1527476418796632

Couldry, N., & Mejias, U. A. (2023). The decolonial turn in data and technology research: What is at stake and where is it heading? *Information Communication and Society, 26*(4), 786–802. DOI: 10.1080/1369118X.2021.1986102

De Siqueira, A. C. (2012). The 2020 World Bank education strategy: Nothing new, or the same old gospel. In *The World Bank and Education* (pp. 67–81). Brill. DOI: 10.1007/978-94-6091-903-9_5

Desai, S., Barik, D., Choudhuri, P., Chouhan, B., & Sharma, O. P., & TIWARI, D. K. (2025). Rethinking Social Safety Nets in a Changing Society: Evidence from the India Human Development Survey. *India Policy Forum : [Papers]. India Policy Forum. Conference, 21*(1), 96–136.

Desai, S., & Dubey, A. (2012). Caste in 21st century India: Competing narratives. *Economic and Political Weekly, 46*(11), 40. PMID: 22736803

Dikeç, M. (2001). Justice and the spatial imagination. *Environment & Planning A, 33*(10), 1785–1805. DOI: 10.1068/a3467

Elwood, S. (2006). Critical issues in participatory GIS: Deconstructions, reconstructions, and new research directions. *Transactions in GIS, 10*(5), 693–708. DOI: 10.1111/j.1467-9671.2006.01023.x

Fang, M. L., Canham, S. L., Battersby, L., Sixsmith, J., Wada, M., & Sixsmith, A. (2019). Exploring privilege in the digital divide: Implications for theory, policy, and practice. *The Gerontologist*, *59*(1), e1–e15. DOI: 10.1093/geront/gny037 PMID: 29750241

Harvey, D. (2010). *Social justice and the city* (Vol. 1). University of Georgia press. DOI: 10.1515/9780820336046

Israel, E., & Frenkel, A. (2018). Social justice and spatial inequality: Toward a conceptual framework. *Progress in Human Geography*, *42*(5), 647–665. DOI: 10.1177/0309132517702969

Iveson, K. (2011). Social or spatial justice? Marcuse and Soja on the right to the city. *City (London, England)*, *15*(2), 250–259. DOI: 10.1080/13604813.2011.568723

Jacobs, M. N., Colacci, A., Corvi, R., Vaccari, M., Aguila, M. C., Corvaro, M., Delrue, N., Desaulniers, D., Ertych, N., Jacobs, A., Luijten, M., Madia, F., Nishikawa, A., Ogawa, K., Ohmori, K., Paparella, M., Sharma, A. K., & Vasseur, P. (2020). Chemical carcinogen safety testing: OECD expert group international consensus on the development of an integrated approach for the testing and assessment of chemical non-genotoxic carcinogens. *Archives of Toxicology*, *94*(8), 2899–2923. DOI: 10.1007/s00204-020-02784-5 PMID: 32594184

Kitchin, R. (2014). Big Data, new epistemologies and paradigm shifts. *Big Data & Society*, *1*(1), 2053951714528481. DOI: 10.1177/2053951714528481

Kitchin, R. (2021). *The Data Revolution: A critical analysis of big data, open data and data infrastructures.*

Kornbluh, F. (2009). Is Work the Only Thing that Pays? The Guaranteed Income and Other Alternative Anti-Poverty Policies in Historical Perspective. *Nw. JL & Soc. Pol'y*, *4*, 61.

Melnyk, L. H., Sommer, H., Kubatko, O. V., Rabe, M., & Fedyna, S. M. (2020). *The economic and social drivers of renewable energy development in OECD countries.*

Monico, J. F. G. (2004). United Nations office for outer space affairs GNSS web-pages. *GPS Solutions, 8*(2), 112–114. DOI: 10.1007/s10291-004-0096-4

Moravec, E. T. (2024). Digital Divide and Social Inclusion: Strategies to Enhance Digital Participation of Marginalized Groups. *Future-Social Sciences and Digital Society, 1*(2).

O'Neil, I., & Ucbasaran, D. (2016). Balancing "what matters to me" with "what matters to them": Exploring the legitimation process of environmental entrepreneurs. *Journal of Business Venturing, 31*(2), 133–152. DOI: 10.1016/j.jbusvent.2015.12.001

Soja, E. (2010). Spatializing the urban, Part I. *City (London, England), 14*(6), 629–635. DOI: 10.1080/13604813.2010.539371

Uwayezu, E., & de Vries, W. T. (2019). Scoping land tenure security for the poor and low-income urban dwellers from a spatial justice lens. *Habitat International, 91*, 102016. DOI: 10.1016/j.habitatint.2019.102016

Van der Linden, M. (2019). The International Labour Organization, 1919–2019: An Appraisal. *Labor, 16*(2), 11–41. DOI: 10.1215/15476715-7323601

Van Dijk, J. (2020). *The digital divide.* John Wiley & Sons.

Van Laar, E., Van Deursen, A. J. A. M., Van Dijk, J. A. G. M., & De Haan, J. (2020). Determinants of 21st-century skills and 21st-century digital skills for workers: A systematic literature review. *SAGE Open, 10*(1), 2158244019900176. DOI: 10.1177/2158244019900176

Villagran de Leon, J. C. (2023). An improved understanding of natural hazards and disasters through the use of satellite technologies: Contributions from un-spider. *Sustainability (Basel), 15*(13), 10624. DOI: 10.3390/su151310624

Viso, M. (2023). ISRO, India. In *Encyclopedia of Astrobiology* (pp. 1558–1559). Springer. DOI: 10.1007/978-3-662-65093-6_1784

Wagle, U. (2002). Rethinking poverty: Definition and measurement. *International Social Science Journal, 54*(171), 155–165. DOI: 10.1111/1468-2451.00366

Xaxa, V. (2011). The status of tribal children in India: A historical perspective. *IHD-UNICEF Working Paper, 7.*

ADDITIONAL READING

Couldry, N., & Mejias, U. A. (2019). *The costs of connection: How data is colonizing human life and appropriating it for capitalism.* Stanford University Press.

Harvey, D. (2010). *Social justice and the city* (Rev. ed.). University of Georgia Press. DOI: 10.1515/9780820336046

Kitchin, R. (2014). The data revolution: Big data, open data, data infrastructures and their consequences. *Sage (Atlanta, Ga.).* Advance online publication. DOI: 10.4135/9781473909472

Lefebvre, H. (1991). *The production of space* (Nicholson-Smith, D., Trans.). Blackwell.

Sen, A. (1999). *Development as freedom.* Oxford University Press.

Soja, E. W. (2010). *Seeking spatial justice.* University of Minnesota Press. DOI: 10.5749/minnesota/9780816666676.001.0001

United Nations Office for Outer Space Affairs. (2023). *Space supporting the Sustainable Development Goals.* https://www.unoosa.org

Van Dijk, J. (2020). *The digital divide.* Polity Press.

KEY TERMS AND DEFINITIONS

Accessibility: The ability of individuals and communities to access and effectively use physical infrastructure, digital technologies, information, and public services. In the context of this chapter, accessibility refers to reducing spatial and technological barriers that prevent the working poor from benefiting from space-based technologies.

Artificial Intelligence (AI): Computer systems capable of performing tasks that typically require human intelligence, including data analysis, prediction, pattern recognition, and decision support. AI enhances the analysis of geospatial data for development planning but also raises concerns regarding bias, transparency, and accountability.

Capability Approach: A development framework proposed by Amartya Sen that defines poverty as the deprivation of people's capabilities and freedoms rather than merely low income. It emphasizes expanding people's opportunities to achieve lives they value.

Digital Divide: The unequal access to digital infrastructure, internet connectivity, digital skills, and technological resources among different social groups. The digital divide limits the participation of marginalized populations in technology-driven development initiatives.

Geographic Information System (GIS): A computer-based system used to collect, manage, analyze, and visualize spatial or geographic data. GIS supports evidence-based planning in agriculture, urban development, disaster management, and poverty alleviation.

Knowledge Integration: The process of combining scientific, technological, indigenous, and local knowledge systems to improve development planning and ensure that technological interventions are socially relevant, culturally appropriate, and sustainable.

Participatory Governance: A governance approach that actively involves communities in planning, decision-making, implementation, and evaluation of development programs to improve accountability, inclusion, and local ownership.

Remote Sensing: The acquisition of information about the Earth's surface through satellite or aerial sensors without direct physical contact. Remote sensing is widely used for environmental monitoring, agriculture, disaster management, and infrastructure planning.

Social Exclusion: A multidimensional process through which individuals or communities are systematically denied access to economic, political, social, and cultural opportunities, leading to persistent inequality and marginalization.

Space-Based Technologies: Technologies derived from satellite systems, including remote sensing, GIS, Global Navigation Satellite Systems (GNSS), satellite communication, and Earth observation, that support development planning, environmental monitoring, disaster management, and poverty reduction.

Spatial Justice: A sociological concept emphasizing the fair and equitable distribution of resources, services, opportunities, and power across geographic spaces. Spatial justice recognizes that social inequalities are often produced and reinforced through spatial arrangements.

Working Poor: Individuals who are engaged in paid or self-employment but remain below the poverty line because of low wages, insecure employment, inadequate social protection, and structural inequalities that limit their economic well-being.

Chapter 2
From Satellites to Solutions:
Leveraging Space Technologies to Uplift the Working Poor

Sovannaroth Chheang
http://orcid.org/0009-0007-3382-7169
Department of Education, Sovanna Elton Foreign Language School (SELS), Phnom Penh, Cambodia

Chheangkhy Hok
http://orcid.org/0009-0000-6431-527X
Department of Humanistic Studies, Indian Institute of Technology (BHU), Varanasi, India

Kimyut Phin
http://orcid.org/0009-0009-5667-7313
Department of English, University of Heng Samrin Thbongkhmum, Tboung Khmum, Cambodia

Sovanna Loch
http://orcid.org/0000-0002-1211-1248
Department of English, Nava Nalanda Mahavihara (Deemed to be) University, Nalanda, India

Sovanna Huot

DOI: 10.4018/979-8-3373-7118-4.ch002

http://orcid.org/0009-0006-8335-442X

Department of Education, National Institute of Education (NIE), Phnom Penh, Cambodia

ABSTRACT

This chapter examines the role of space technologies as potential tools for the development of the poor, especially the working poor, who continue to be excluded from the development process. It brings together theory and practice, to show how remote sensing, satellite communication, geospatial mapping and Earth observation technologies, can be integrated to improve agricultural productivity, increase access to education and healthcare, create jobs, and foster economic and social empowerment. The chapter also addresses the socio-technological, financial, and policy-related barriers to the use of the identified space technologies and suggests that these be implemented through multi-stakeholder partnerships, collaboration, and inclusive governance. From a high-tech to a low-tech perspective, this chapter views space technology as a strategy that fosters innovation and equity. It proposes that the linking of space technologies with terrestrial applications can help to diminish disparities and enhance sustainable development.

I. INTRODUCTION

Poverty remains one of the most formidable challenges faced by humanity, spanning economic, social, and geographical boundaries. Millions of individuals endure deplorable living conditions. Escaping such circumstances is a complex endeavor. Over 700 million people live below the poverty line, of whom nearly 350 million are classified as working poor—economic contributors unable to earn sufficient income to meet their basic needs (Worl Bank, 2022). This describes the reality of almost half of the economically active population. Sen (2015) stresses that poverty is not merely a lack of money, but a deprivation of basic freedoms and the capacity to live a full and meaningful life. This reflects that those below the poverty line are not just economically inactive but also miss out on social privileges enjoyed by active members.

Scholars such as Hickel (2017) and Milanovic (2019) advanced the argument that poverty is a consequence of structural inequity that permeates the world economic system. In the United States, the rural counties with chronic poverty that have poverty rates of over 20 percent for several decades illustrate the phenomenon. These counties have poor physical and social infrastructure and are characterized by low levels of education and racial inequities (U.S. Census Bureau, 2023). The situation is even more complicated globally. Sumner (2016) notes that the global poor are now primarily found in middle-income countries, which are characterized by social inequity, exclusion, and economic marginality. Thus, as Devereux and Sabates-Wheeler (2004) observe, this poverty is understood as dynamic and relational, shaped by cycles of vulnerability, labor market instability, and environmental conditions. The "working poor" embody this paradox: they contribute labor and productivity to the economy, yet still do not achieve economic stability. For the last 50 years, poverty alleviation has employed strategies such as cash transfers, microfinance, social safety nets, Agricultural Subsidies, and educational programs. Though important outcomes have been achieved, most have failed to eradicate poverty or address its causes. Banerjee and Duflo (2013) posit that small, evidence-based interventions that are theory-driven can improve welfare, but the authors acknowledge such initiatives can be contextually bound and have limited scalability.

Similarly, Sachs (2015) advocates integrated development interventions while acknowledging that many such initiatives are prone to poor coordination, corruption, and logistical lapses. Several recent works have criticized the conventional strategies for poverty alleviation for being passive, or rather, not being proactive. Ravallion (2016) argues that most approaches ignore systemic inequity and existing knowledge gaps. A significant paradox is that there are major investments in education and health care, and yet many rural areas are still underserved, and this can be attributed to poor geographical coverage, weak data collection, and ineffective monitoring (Dercon, 2022). Consequently, scholars such as McGee and Edwards (2016) have drawn attention to the so-called accountability gap in development: programs fail because decision-makers are uninformed or poorly informed, and the information they lack is timely, accurate, and of relevant spatial granularity.

The shortcomings of these outdated approaches have prompted a so-called "innovation imperative" in poverty-related research and policy. Technological transformation likely can alter deeply embedded inequalities and, when properly directed, may be inclusive. The potential of space technology as a transformational tool for innovation, as detailed by Sandau (2010), may help circumvent physical and institutional barriers. It provides innovative tools for real-time data collection, remote monitoring, and communication. The innovative intersection of digital and space technologies is, as Sandau (2010) suggests, likely the most useful technology available to governments and NGOs for providing the detailed evidence needed to develop responsive and adaptive social protection systems. The first couple of decades of the 21st century have been marked by an unprecedented revolution in the accessibility and affordability of space technology.

Proliferation of small satellites, commercially available geospatial data, and innovations by private industry have largely democratized what was previously reserved for state space agencies. Scholars such as Johnson-Freese (2016) and Krige (2019) note that the "New Space" era is transforming telecommunications and navigation, while also fundamentally changing how societies collect and analyze data about Earth systems. The potential for such technologies to advance development is increasingly acknowledged in policy and academic circles. The satellites used for Earth observation and remote sensing are invaluable for assessing and mapping poverty, monitoring the state's agriculture, assessing environmental degradation, and forecasting natural disasters. While Jean et al. (2016) used machine learning to correlate satellite images to poverty at granular levels, Henderson et al. (2012) were the first to utilize nighttime satellite images as proxies for economic activity.

More recently, Kim et al. (2022) show that the fusion of high-resolution satellite images and artificial intelligence (AI) enables the near-real-time assessment of poverty. Thus, governments optimize their resource allocation. These studies show that satellite data can enhance even the most targeted and precise policies by providing policymakers with empirical evidence. Beyond observing the Earth, satellite communications help close the digital divide. In broadband-scarce and broadband-unavailable areas (e.g., remote or conflict-affected regions) where terrestrial networks do not exist or are not dependable, satellite broadband supports online education, telehealth, and e-commerce, as well as e-governance (Maral & Bousquet, 2020). Satellite broadband has the potential to support the International Telecommunication Union's (ITU, 2023) advocacy that satellite-enabled broadband is essential for digital inclusion and to help achieve the UN Sustainable Development Goals. Some scholars, such as Castells (2010) argue that access to information networks constitutes a new form of social capital, so connectivity is not merely technical infrastructure; it is a condition for modern citizens and their participation.

Despite the optimistic perspective, there are critiques. Phillips (2017) warns that technological systems, even space-based ones, can replicate existing power structures. If access and governance are inequitable, the systems perpetuate those injustices. Mulhaney et al. (2024) argue that technology must be socially embedded. Without the active participation of the community and the local governing structures, even the most powerful technology is more likely to become a tool of exclusion rather than empowerment.

Considering the stated challenges, this chapter has three objectives that will be addressed in a systematic manner:

1. Analyzing space technology as a means of reducing poverty, more specifically, the data, connectivity, and visibility it can provide to marginalized groups.
2. Assessing practical uses of these technologies for the working poor, such as in agriculture, healthcare, education, financial inclusion, and disaster management.
3. Assessing the institutional, policy, and partnership frameworks that enable the conversion of satellite data into actionable developmental outcomes and identifying pathways that drive social impact from technological innovation.

The objectives of the chapter, then, span the conceptual and practical domains of aligning space-enabled instruments with strategies to reduce poverty and configuring innovation ecosystems to support the inclusive distribution of benefits.

II. CONCEPTUAL AND THEORETICAL FOUNDATIONS

2.1 Defining the Working Poor

The intersection of employment and poverty research is critical for defining the working poor. The working poor, as a phenomenon, is understood to be individuals who perform paid work, but who remain in a state of poverty as defined by a given country and the international community (Bureau of Labor Statistics, 2024). Employment as a form of poverty alleviation has been critiqued by Rodgers (2018), who argues that working poor people expose the employment paradox. He explains that employment does not equate to a living wage, adequate protection, or social protection, especially in informal and precarious employment situations.

The Fifth International Labor Organization (ILO) estimates say nearly one-fifth of the world's workers are working poor, partially due to structural flaws in the labor market (ILO, 2022). According to Fields (2012), agriculture, informal services, and small-scale manufacturing are all low-productivity sectors that entrap workers in underemployment, stagnant skills, and income. Women and youth are the predominant victims (Kabeer, 2015), occupying the most vulnerable, poorly paid, and part-time jobs. Systemic barriers aggravate these risks. Workers in informal labor systems are often excluded from social protection, health coverage, and pension systems, locking entire families in poverty for generations (Lund & Nicholson, 2003). Geographic isolation makes these inequities worse. In developing countries, the rural population often lacks basic education, infrastructure, and markets, sustaining the low-wage labor cycle and reducing social mobility (Dercon, 2022). Further, the lack of adequate and effective use of information and communication technologies (digital exclusion) acts as a new barrier to the working poor in the digital age (Helsper, 2021).

2.2 Technology and Development Frameworks

The relationship between technology and development has been driven by two opposing frameworks. The first is techno-optimism, which views technology as a positive phenomenon; in contrast, critical frameworks examine technology in relation to its social and political consequences. The modernization theory of Rostow (1960) is an example of the first wave of techno-optimism, where it is argued that societies move through a series of defined stages of economic development because of technological change (Rostow, 2000). In more recent times, scholars such as Friedman (2006) maintained this optimism when he argued that globalization and digital interconnectivity would level the economic playing field.

Nevertheless, critical theorists tend to argue against these assumptions. Heeks (2008) argues that digital development, in most cases, has little neutrality and reproduces existing power relations in the absence of intentional, inclusive design. Warschauer (2003) argues that owning technological tools does not guarantee empowerment and introduces the term digital capital. Empowering financially, he argues, is achievable only through a combination of tools, local and institutional partner capacity, and contexts. Equally, Avgerou (2010) argues that in the absence of social and governance frameworks, irrationality exacerbates social injustices. This imbalance, when considered alongside optimism and critique, provides a basis for viewing the study of technology as a social system rather than a mere instrument. Development is not about the provision of tools; it is about people's capabilities, as Sen (2015) states.

2.3 Space Technologies in Context

With the development of space-based technologies, new approaches for observing and understanding the planet have emerged. Remote sensing, Global Positioning System (GPS), communication satellites, and Earth observation (EO) systems have traditionally been used for defense, science, or commercial applications. Because of what Krige (2019) and Johnson-Freese (2016) refer to as the democratization of space, now even low- and middle-income countries can utilize these technologies for civilian and development purposes.

EO satellites operate 24/7, capturing high-resolution images and delivering hyperlocal data on land use, vegetation, water, and urban development. Henderson et al. (2012) explained that light emissions at night have a strong correlation to economic activity and can be used for low-data regions as a proxy for economic activity. Jean et al. (2016) used machine learning to create low-cost, high-speed granular poverty maps that surpassed the performance of traditional surveys. More recently, Kim et al. (2022) showed how the combination of Artificial Intelligence (AI) and EO data can be used to monitor welfare changes in near real-time. This allows for responsive policies to be implemented. Just like the systems that provide coverage to the most inaccessible locations are built on the foundation of GPS and communication satellites, Maral and Bousquet (2020) describe the importance of satellite telecommunications in telemedicine, remote education, and the coordination of services during crises, which are all important for underserved populations. Also, combining earth observation (EO) data with ground-based Internet of Things (IoT) devices and cloud computing enables smallholder farmers to use predictive analytics to know in advance the changes in the weather and to make better use of the resources (Aggarwal et al., 2019).

All these changes highlight how space-based systems have transformed from mere technological artifacts into integral constituents of an emerging socio-technical ecosystem. Yet, Phillips (2017) argues that the current geography of access to space-derived information remains uneven, with most control held by high-income countries and large, monopolistic corporations.

2.4 Sustainable Development Lens

This chapter also looks to the UN Sustainable Development Goals (SDGs) for guidance. The SDGs outline the vision to end poverty in all forms and ensure equitable growth. Among these, SDG 1 aims to end extreme poverty by the year 2030, while SDG 9 focuses on Industry, Innovation, and Infrastructure, specifically on the creation of inclusive and safe technological advancement and resilient infrastructure (Sayed, 2015; United Nations, 2015). The SDGs, in their entirety, advocate improving society through technology, rather than advancing technology for its own sake.

According to the UN Office for Outer Space Affairs (UNOOSA, 2024), most SDGs, including those relating to climate, food security, education, and healthcare, significantly benefit from space technology. More specifically, space-based data contributes to evidence-based decision-making and identifies the location and distribution of poverty, ecosystem loss, and disaster risk. Brunswick (2025) and Tennin and Brunswick (2024) assert that merging the space development tools and sustainable development frameworks requires collaboration across various fields, integrating the technical capacity of the space development domain and the social development focus of the development domain. When viewed through a sustainable development lens, space technologies can be considered a means of enacting inclusive innovations. Enhancing agricultural productivity (SDG 2), resilient infrastructure (SDG 9), climate adaptation (SDG 13), and equitable access to education and health (SDG 4 and SDG 3) are all within the scope of space tech. However, as Sachs (2015) cautions, the potential of such benefits depends on effective governance and equitable resource distribution. Space technologies, like many other innovations, will likely reproduce the negative development spaces of the past, as shown in Figure 1.

Figure 1. Conceptual pathways linking space-based technologies to inclusive development outcomes

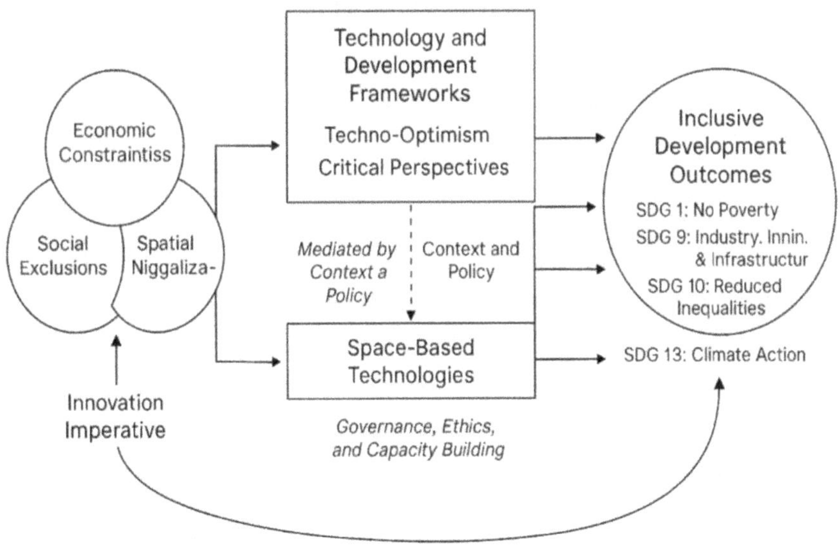

III. SPACE-BASED TECHNOLOGIES: TOOLS AND APPLICATIONS

3.1 Monitoring Agriculture, Urbanization, and Environmental Change

Henderson et al. (2012) were the first to show that nighttime satellite imagery reveals economic activity and captures areas with missing or poor economic data. Whereas Jean et al. (2016) used daytime satellite images and machine learning to predict poverty in Sub-Saharan Africa with accuracy comparable to household surveys. Jean et al. (2016) demonstrated the ability of remote sensing to fill data gaps that obstruct effective targeting of poverty. Remote sensing (Aggarwal et al., 2019) also provides real-time data on crop conditions, soil moisture, and vegetation, as do NASA's MODIS and ESA's Sentinel in agricultural settings (Meroni et al., 2019). Such data allow smallholder farmers, who constitute a significant segment of the working poor, to improve their planting calendars, conserve water more effectively, and adapt to climate variability. Becker-Reshef et al. (2010) underscore the importance of satellite crop forecast data in planning for national food security, as it enables timely interventions in vulnerable areas.

EO technologies help monitor the growth of informal settlements, the shortfall of various urban infrastructures, and the impact of urban sprawl on the environment. As Dare and Fraser (2001) state, high-resolution images have the potential to map informal settlements and peri-urban regions that remain unrecorded in official counts and, therefore, help plan and allocate urban services. Remote sensing of Earth helps monitor and manage the use and allocation of ecosystem services. As Thornton et al. (2014) and Greene (2018) state, climate change disproportionately affects the working poor, as their livelihoods are labor-based and depend on direct interaction with ecosystems affected by weather and climate. By utilizing remote sensing data and combining it with socioeconomic data, governments and NGOs can anticipate and prepare for crises, rather than merely respond to them, which serves to empower individuals, as stated by Sen (2015). The European Organization for satellite observation (EO) and satellite communication systems (SatCom) offer different services, with the former focusing on observation and the latter on communication. Castells (2010) recognizes communication and access to information as critical factors for social and economic participation.

3.2 Increasing Availability of Education, Telehealth, and Market Access

Before the introduction of satellite broadband, participants in online learning and telehealth services, as well as smallholder farmers, faced significant barriers, with access being the most prominent. Maral and Bousquet (2020) state that both geostationary and low-earth orbit (LEO) satellites can provide access to high-speed networks in situations where traditional networks cannot. For example, in the education sector, UNESCO's satellite-based distance education program, "Education for All via Satellite," provides remote areas with access to online resources, including digital curricula and teacher training (UNESCO, 2023). Similarly, satellite telemedicine in Sub-Saharan Africa and South Asia has linked community health workers and urban hospitals to provide real-time diagnostics and consultation (WHO, 2025).

Regardless of how big the digital world grows, rural areas can remain even less connected. Most of the 2.6 billion people without internet access live in developing areas of the world, according to the International Telecommunication Union (ITU, 2023). Rapidly evolving satellite communications technologies, especially those utilizing Low Earth Orbit (LEO) constellations, are of paramount importance to achieving global digital inclusion. Maîtrot et al. (2021) insist that 'connectivity is the new infrastructure of equality' because without it, the marginalized are left out of civic engagement, financial inclusion, and work opportunities. Although it is a critical first step, simply providing technology is not enough. As Toyama (2015) and (2017) states, without the necessary human and organizational resources, technology amplifies the existing inequalities.

Hence, the use of satellite communication must be combined with initiatives to improve digital skills and local empowerment to ensure it is an inclusive rather than exclusionary tool. Geospatial technologies, such as GPS, Geographic Information Systems (GIS), and spatial analytics, provide the means to analyze and visualize socioeconomic data over time and space. Poverty spreads unevenly and often concentrates in spatial hotspots, which are regions characterized by environmental fragility, poor infrastructure, and weak institutional frameworks (Kanbur & Venables, 2005). Geospatial mapping helps both visualize and quantify these concentrations. For example, Chen (2019) demonstrate how the combination of satellite imagery and household surveys leads to the creation of detailed poverty maps, which show disparities at intra-regional levels that are often overlooked in national averages.

In addition, GPS-based surveys help collect and spatially tag data on employment, health, and education. This collection helps link macro trends to micro realities. As Mellor (2017) puts it, spatial analytics transforms poverty from an abstract statistic into a mappable, actionable phenomenon. Such a transformation fosters ease of use, accountability, and the design of focused policies.

3.3 Planning Based on Evidence for Policymakers and NGOs

There is a growing reliance by governments and NGOs on geospatial data for program planning and subsequent monitoring. The United Nations Global Geospatial Information Management (UN-GGIM) initiative provides guidance on integrating national statistics with geospatial data frameworks. Using these systems, development agencies can identify locations where the lack of infrastructure coincides with high levels of poverty or where environmental threats to livelihoods are present. Brunswick (2025) makes clear that evidence-based interventions, as opposed to political, anecdotal, or other forms of non-evidence-based interventions, shift governance from a political process to a more operational one.

Satellites provide monitoring and early warning of floods, droughts, and storms, and enable observation of weather and climate changes. The Global Disaster Alert and Coordination System (GDACS) offers and integrates Earth Observation (EO) data with terrestrial sensors and systems to facilitate immediate communication, thereby delivering real-time alerts to governments and communities worldwide, according to the United Nations Office for Disaster Risk Reduction (UNDRR, 2023). During the recovery phase, remote sensing helps with damage evaluation and redesigning aid distribution. Kerle (2013) illustrates how remote sensing after a disaster helps sustain efficient resource allocation and monitor rebuilding in inaccessible regions. Furthermore, when ground-based infrastructure is deficient, satellite communication (SatCom) enables remote operational support from the control center, thereby facilitating coordination among humanitarian actors (Maral & Bousquet, 2020). From a climate change perspective, the working poor and the intersection of poverty and disaster underscore the need for integrated response systems. The working poor, particularly in agriculture or coastal livelihoods, are heavily affected by flooding, drought, and extreme heat. It is advocated that, where remote sensing (RS) is integrated with socio-economic data, governments develop anticipatory social protection systems that are triggered by disasters, and with no harm in pre-emptive disaster systems, aid is given (Hallegatte et al., 2020).

In summary, these fields show space-based technology as a versatile development tool. Remote sensing offers data for planning; satellite communications improve connectivity; geospatial analysis and cloud data support governance; and disaster planning boosts system resilience. When integrated into the outlines of a fair policy framework, these technologies serve as an observation system while empowering the working poor. Repurposing tools of science for the pursuit of social justice, not simply the act of invention, is what Sachs (2015) refers to as the technological frontier of development. As shown in Table 1, space-based systems, when integrated properly, exemplify this potential.

Table 1. Space-based Technologies: tools and applications

Category	Primary Functions	Applications	Key Benefits
Remote Sensing & Earth Observation	Satellite imaging; environmental and agricultural monitoring; change detection.	Monitoring crop yields, urban expansion, deforestation, and resource management.	Improved food security, sustainable farming, early warning for environmental risks.
Satellite Communication Systems	Transmission of data, voice, and broadband signals via satellites.	E-learning, telehealth, digital markets, and mobile financial services in remote areas.	Expanded access to education, healthcare, and income opportunities; reduced digital divide.
Geospatial Data & Mapping	Spatial analysis using GPS, GIS, and Earth observation datasets.	Identifying poverty hotspots, infrastructure planning, and resource allocation.	Evidence-based policymaking, targeted development programs, enhanced transparency.
Disaster Management Applications	Real-time monitoring of weather, floods, droughts, and other hazards.	Early warning systems, post-disaster assessment, and recovery coordination.	Protection of livelihoods, reduced vulnerability to climate shocks, faster relief delivery.

IV. SOCIOECONOMIC BENEFITS FOR THE WORKING POOR

Future space-based technologies could significantly benefit the working poor by enhancing productivity, access, employment, financial capabilities, economic engagement, and inclusion. In developing systems-integrated space-enabled tools, as noted by Brunswick (2025), along with Sachs (2015), claim such tools create a shift in focus from helping the poor to helping them use their capabilities. As a primary source of income in the developing world, farming supports the livelihoods of almost all poor working people worldwide. Smallholder farmers (FAO, 2022), Those with farms under two hectares are key food producers, contributing over a third of global food. They face barriers like low productivity, climate issues, and underdeveloped space tech, which reduce market access. Remote sensing and Earth observation (EO) are starting to address these challenges. Meroni et al. (2019) and Aggarwal et al. (2019) state that farmers can use satellite-based weather forecasting systems to understand rainfall, drought, and temperature, as these systems provide important, precise, localized data. EO systems can capture images at high frequencies, unlike traditional meteorological systems, thereby enabling near-immediate decision-making. This is especially important in climate-vulnerable areas like Sub-Saharan Africa and South Asia, where most of the agriculture is rain-dependent (Greene, 2018; Thornton et al., 2014).

Becker-Reshef et al. (2010) illustrate the extent of EO-based soil analytics where moisture and fertility data sourced from satellites are reshaping 'smallholders' of managing land and water. EO analytics show the restoration of optimized irrigation and fertilizer use, which reduces input costs and ecological impact. As Sandau (2010) maintains, such innovations transform traditional farming into a data-driven livelihood, empowering farmers with knowledge once limited to industrial agriculture. Monitoring crops in real time helps people in the supply chain achieve higher profits. When smallholders get formal contracts, insurance, and credit, they use EO data for yield and crop quality verification (Brickell et al., 2020). According to Dare and Fraser (2001), the use of satellite data in agribusinesses improves trust and the operational efficiency of the value chain. Therefore, the use of satellite technology in agriculture aligns with Sen's (2015) construction of capabilities for development, reinforcing the possibility of people going beyond mere subsistence to economic activity. UNESCO (2023) points out that satellite-supported e-learning systems enable remote communities to access educational resources and teacher training without relying on terrestrial networks. Maral and Bousquet (2020) state that geostationary satellites offer uninterrupted broadcasting support for distance education programs. This results in uninterrupted learning, even during disasters and network outages. For example, satellite connections in sub-Saharan Africa and in parts of Southeast Asia have made virtual classrooms possible, where traditional schools could not be built (Maîtrot et al., 2021).

In addition, satellite broadband and digital literacy initiatives strengthen the cognitive and operational skills needed to engage in the digital economy (Helsper, 2021). Equally, as Brunswick (2025) states, when people are connected, education becomes more than a privilege; it becomes a right. This enables the working poor to access jobs and businesses that require higher skills. The delivery of health services also benefits significantly from satellite-based systems. Remote consultations, diagnostics, and Health monitoring can be done via telemedicine services, supported by satellites, the World Health Organization (WHO, 2025). The African Telehealth Network, for example, offers satellite-based telehealth services in several countries and integrates them in countries that do not have hospitals. EO data also help in tracking the geographical distribution of certain diseases and monitoring the environmental conditions related to the spread of malaria, cholera, and dengue (Kim et al., 2022). Such systems are designed to minimize travel time and costs for low-income users. As noted by Toyama (2015), technology amplifies intention: with a social platform, satellite communication can improve access to services in a way that digital inclusion translates to real gains in health and wellness.

The spread of space-related technologies is creating a range of jobs beyond high-tech industries. According to scholars Johnson-Freese (2016) and Krige (2019), the global space economy, which is expected to exceed $1 trillion by 2040, is benefiting not only engineers and scientists but also creating jobs for technicians, analysts, and entrepreneurs in adjacent industries. The increasing availability of satellite data and the adoption of geospatial technologies are driving demand for digital professionals, data scientists, and GIS technicians in the Global South (Mulhaney et al., 2024). International organizations, including the United Nations Platform for Space-based Information for Disaster Management and Emergency Response (UN-SPIDER), are sponsoring training programs to improve developing countries' ability to use satellite data. Such programs are for the working youth, especially the poor, to equip them with digital competencies required by the Fourth Industrial Revolution (Schwab, 2016).

Beyond formal employment, space technologies also enable small-scale entrepreneurship. Local innovators use satellite data to develop weather apps, agricultural advisory services, and mapping solutions tailored to community needs (Sandau, 2010). As Brynjolfsson and McAfee (2014) argue, digital transformation thrives when technological infrastructure catalyzes local creativity rather than merely displacing existing labor. By linking space data with local business models, working populations can transition from consumers to producers of digital value. Digital upskilling is integral to ensuring equitable participation in the emerging space economy. Brunswick (2025) and Tennin and Brunswick (2024) maintain that inclusive space education programs, particularly those targeting women and youth, are essential for diversifying the space workforce and extending benefits to marginalized groups. Financial exclusion remains a defining feature of poverty. The World Bank's Global Findex Database (2022) reports that over 1.4 billion adults worldwide remain unbanked, with most residing in rural or remote regions. Satellite connectivity bridges this gap by enabling mobile banking, digital payments, and microfinance services in areas beyond terrestrial network coverage. Satellite-linked mobile networks enable low-income individuals to conduct banking transactions without access to physical infrastructure. Brickell et al. (2020) observe that such systems underpin platforms like M-PESA in East Africa, where mobile money has revolutionized savings, remittances, and access to credit.

Moreover, geospatial data enhances the accuracy of credit scoring and insurance modeling by linking household or agricultural data to spatial indicators of risk (Chen, 2019). This allows microfinance institutions to extend credit to informal workers who lack conventional financial records, expanding economic participation. Microfinance institutions (MFIs) increasingly utilize satellite-enabled data to assess loan performance and disaster exposure. Hallegatte et al. (2020) note that integrating EO data with microfinance analytics can trigger insurance payouts or credit extensions when environmental shocks threaten livelihoods.

In this sense, financial inclusion becomes intertwined with resilience: satellite-enabled systems not only provide access to capital but also protect it from climate risks. As Sachs (2015) asserts that financial empowerment is most sustainable when underpinned by technological systems that anticipate vulnerability rather than merely compensate for loss. However, as Sen (2015) reminds us, technology alone cannot ensure freedom; it must be accompanied by equitable policies, education, and governance. Satellites offer more than data; they can be integrated into social systems that promote inclusion and empowerment. Aligning innovation with justice, space-enabled development could reshape technology's role in reducing poverty, as summarized in Table 2.

Table 2. Socioeconomic benefits of space-based technologies for the working poor

Domain	Mechanisms	Functions	Impact
Agricultural Productivity	Remote sensing, Earth observation, weather satellites, soil moisture analysis	Crop monitoring, yield forecasting, drought detection, precision agriculture	Improved food security, higher yields, reduced input costs, climate resilience for smallholders
Access to Services	Satellite communication systems, broadband connectivity, tele-education, telemedicine	Distance learning, teacher training, virtual classrooms, remote diagnostics and health consultations	Expanded access to education and healthcare; empowerment through knowledge and well-being
Employment Opportunities	Satellite data processing, GIS applications, space-enabled entrepreneurship, digital skills training	Jobs in geospatial analysis, local startups using satellite data, community-level innovation hubs	Increased employment and income generation; digital skill enhancement for marginalized workers
Financial Inclusion	Satellite-enabled mobile networks, geospatial credit assessment, microfinance analytics	Mobile banking, digital payments, satellite-based insurance and credit scoring	Access to financial services, risk reduction, enhanced economic resilience for unbanked populations

V. CHALLENGES AND BARRIERS

The potential benefits of space-based technologies in alleviating poverty and empowering the working poor are numerous; however, many obstacles to their adoption and equitable use remain. Heeks (2008) and Sachs (2015) argue that the first order of business in such unbalanced arrangements that exhibit technological innovation is the policies aimed at reducing inequalities. Integrating such technologies to alleviate poverty among the working poor will require removing numerous structural, economic, and sociocultural obstacles that lead to limited accessibility, heightened suspicion, and trust issues. Investments in space-based technologies are highly capital-intensive. Technologically, the costs associated with satellite launches and the accessibility of satellite data have drastically diminished; however, inadequate data accessibility and infrastructure for low- and middle-income countries (Maral & Bousquet, 2020). Most developing nations still face prohibitive costs that keep them from obtaining the space capabilities that Johnson-Freese (2016) describes as including ground stations, data processing facilities, and the necessary technical expertise.

Accessibility, in terms of cost, is not an issue for all satellite data, as a spectrum of data processing, storage, and analysis costs is incurred with different satellite data, including NASA's Landsat and ESA's Sentinel data (Kim et al., 2022). Even at the local and regional levels, translating raw satellite data into useful, actionable insights is a process that many local governmental agencies lack the computing capacity to perform. Brunswick (2025) makes clear that until precise goals of data analysis are established and aimed at developing local capacities, the hope of democratized data accessibility is, at best, a hope. It is also a question of the regional and local governmental agencies' ability to provide the most basic supporting resources. These resources are reliable and provide electricity, data, and high-speed Internet access. These resources are also needed to fully assess the potential that satellite data services can provide (Maîtrot et al., 2021). Providing technology in areas with the greatest concentration of poverty, which are most often remote and rural, is not new. The underdevelopment of the required basic tools and technologies is. The distribution of technology and its advances is fundamentally inequitable (Krige, 2019).

The greater the latitude and complexity of the data required to analyze satellite data, the greater the reliance on an outside expert to provide it. Mulhaney et al. (2024) caution that developing countries lack localized analytical capacity and that the remote satellites that collect the data on which the developing nations depend are becoming data consumers instead of becoming data producers. The economic factors influencing the adoption of space-enabled solutions are a perfect example of the digital inequality paradox: when technology becomes available, those who are most economically vulnerable often stand to gain the most, yet are also the least able to afford it. Brickell et al. (2020) explain that although remote satellite coverage is a potential solution to the digital divide by providing rural users with new access to economically relevant connections, the coverage remains unaffordable to most low-income users, particularly in areas where terrestrial networks do not exist.

Satellite broadband remains economically inaccessible to rural users, primarily due to the subscription fees, the cost of the satellite equipment, and the data usage fees (ITU, 2023). Furthermore, the limited number of broadband satellite service providers and the insufficient competition in rural areas further intensify the affordability gap. Toyama (2015) rightly argues that in low-resource environments where both financial and human capital are in short supply, high-cost technological innovations paradoxically increase the level of exclusion the innovations were designed to address. In their work, Brynjolfsson and McAfee (2014) mention that investment incentives are weak in poor and rural areas, which also contributes to high levels of poverty and a lack of investment in developing new technologies. In these empty spaces of innovation, the only things to innovate are the weak socioeconomic conditions that blend finance and subsidies frameworks struggle to change. To create inclusive innovations, innovation needs to economically 'de-risk' these areas. Technical and economic factors are not the only things that create barriers to the adoption and continued utilization of technology. Warschauer (2003) indicates that the digital divide is not only about access to technology, but also about understanding, trust, the daily integration of that technology, and its overall meaningful use.

Deficits in digital literacy are a constant impediment. Many working poor in rural and marginalized areas lack even the most basic skills to use digital devices and to understand data and information based on satellite technology (Helsper, 2021). The presence of technological infrastructure does not change the second-level digital divide (Hargittai, 2003), a phenomenon that inhibits the effective use of access to the information and data that the technology provides. Resources are not empowering in and of themselves; as Kabeer (2015) states, they require effective utilization to be empowering. Trust and perceptions regarding culture play key roles. Some communities may be skeptical of satellite monitoring because it is viewed as foreign surveillance (Phillips, 2017). This kind of skepticism may result in the neglect of satellite-based EO and geo-programs, especially where communities are excluded from the design and decision-making process. Inclusion means partnerships and collaboration, as Brunswick (2025) notes, stressing the importance of working with local actors for the innovations to be relevant and legitimate. The impact of social barriers is compounded by gender inequalities. Women, many of whom are part of the informal economy, are often sidelined from the use of digital technologies because of cultural attitudes, caregiving roles, and safety concerns (Kabeer, 2015). Programs that are space-enabled and lack deliberate gender-sensitive approaches will most likely reproduce rather than ameliorate existing inequalities. The integration of space technologies into the poverty alleviation process is possible with adequate governance systems, a deficient area in the Global South. Both Sachs (2015) and Sen (2015) point out that the absence of governance of technology use may lead to unfair and undesirable outcomes.

Technological advancement is fast-paced; space governance is the opposite. Most countries lack space policies, space governance regulatory bodies, or data-sharing agreements (UNOOSA, 2024). This lack creates silos and poor collaboration among public sectors, non-profits, and the private sector. Kim et al. (2022) clarify that the division of roles among the common, science, and ministries creates bureaucratic stagnation; therefore, this study has postponed or canceled initiatives. Innovation policies have focused on high technologies and defense. They bypass poverty mitigation, and the innovation that Avgerou (2010) argues is inclusive and aimed at closing the technological and social divides is not being practiced. Without such an aim, users of state funding perpetuate an exclusive, elite research focus on local issues. He then suggests that social impact assessments be incorporated into space policy to provide a basis for a space-for-development innovation. Governance of data is a growing concern, particularly the collection of data by satellite technology, spatial technology, and the analytics of such technologies. Data, in the form of satellite pictures and spatial analytical technologies, can provide sensitive information about a person or a group. Heeks (2008) suggests that the absence of stringent data protection in developing countries can, and has, harmed the people who are least protected by surveillance and other exploitative mechanisms. The use of technology should be accompanied by ethical frameworks and community consent procedures to enhance a community's trust and to protect the community's rights.

Finally, space-based innovation is impacted by the same geopolitical and economic disparities that influence global development. A select few countries and corporations own space infrastructure, satellites, launch vehicles, and the digital networks that interconnect them. As Phillips

(2017) notes, ownership of space-based information and economic power corresponds to inequities on the ground, which he terms the cosmic divide. Third-world nations are often reliant on the data and services provided by international organizations or private companies, resulting in what Krige (2019) describes as dependence asymmetries. Despite increased accessibility from open data initiatives, critical dependencies remain in satellite ownership, frequency licensing, and cloud processing. Sachs (2015) argues that these dependencies strengthen the core-periphery division of the global economy, in which the benefits of innovation are disproportionately distributed to those who control production and data flows.

As previously mentioned, the obstacles to the fair investment of space-based technologies are as technical as they are systemic. Space-based technologies can be economically exclusive and socially distrustful, and policy vacuums will remain. Sachs (2015)and Sen (2015) highlight the importance of an innovative and equitable social order for sustainable advancement. The removal of inequities, along with united international collaboration, multi-level governance, and capacity building, is required to address inequities. Only when the social order is inclusive and the policies are ethical and equitable will space technologies be able to empower the economically vulnerable and the socially marginalized.

VI. PATHWAYS FROM SATELLITES TO SOLUTIONS

The potential of space-based technologies to alleviate poverty depends on a distinct ecosystemic approach that integrates innovation, governance, and inclusion. The previous section highlighted key challenges, whether technological, economic, social, or policy-related. In contrast, this section outlines the most feasible pathways to transform operational satellites into tools that benefit society. As Brunswick (2025) states, the transition from satellites to solutions is determined not solely by technology but also by the connections among the network of stakeholders, institutions, and communities, and by space-based data for development. Interdisciplinary work is defined by the integration of two or more disciplines. In this case, multidisciplinary work is needed to address space-based poverty-alleviation initiatives. There is a need to combine the efforts of governments, the private sector, civil society, academia, and community organizations. Sachs (2015) and Sen (2015) suggested that sustainable development results from the fusion of technological capabilities and active participation by governments and the people. While the private sector contributes innovation, efficiency, and the ability to scale solutions to problems, the government is the primary actor in designing control mechanisms and default financing structures. Because of this, Johnson-Freese (2016) describes the collaboration of national space agencies and commercial providers as a means of widening the availability of commercially affordable data in the Global South.

International organizations and NGOs are the primary intermediaries between technology providers and final users. UNOOSA (2024) uses a technology-pedagogy approach to the Space4Women and Space for SDGs initiatives, which seek to promote the integration of women and underserved populations in the benefits of space. These partnerships, as Sandau (2010) points out, build a cross-sectoral value chain; every partner in the collaboration contributes a unique, valuable, and distinct resource, whether it is tech, policy, or community engagement. It is essential that local communities are not just recipients of benefits but also co-designers of space-enabled solutions. Toyama (2015) stresses that local adoption is vital for any technology to succeed; communities must be involved in problem identification, defining their data needs, and tailoring solutions to their sociocultural milieu. Capacity building is integral to the sustained diffusion of technology. According to Heeks (2008), digital inclusion entails more than just having the requisite hardware; end users should also be able to wield, understand, and ideate on the technology. For the working poor, learning digital, analytical, and adaptive skills to endure the changes of the space-tech ecosystem is equally important. The training programs offered by individuals and institutions in remote sensing, GIS, and satellite data analysis enable the localization of technology for development. UN-SPIDER has equipped thousands with data fusion and disaster mapping skills (UNOOSA, 2024). Projects such as ISRO's Bhuvan in India, which offers free satellite data and online courses to local researchers and NGOs, are also national-level initiatives.

Revamping vocational training is critical to grassroots capacity building. Brunswick (2025) asserts that teaching the working poor hands-on technological literacy is a means to elevate their employability and civic engagement. Data-driven farming is the kind of technology that lessens the impact of climate change shocks on farming (Aggarwal et al., 2019). Farmer students of satellite-based weather and soil systems can make better agricultural decisions. Building capacity also applies to institutions. NGOs and government agencies aiming to achieve effective poverty alleviation need the capacity to analyze satellite data and integrate it into their policymaking. As Krige (2019) says, international fellowships, data-sharing hubs, and open-source technologies are examples of knowledge transfer that enable developing countries to innovate, rather than just receive, in the global space ecosystem. The potential of technology is most fruitfully realized through policy. Sachs (2015) argues that the new must always be integrated into frameworks of social justice, which guarantee equal participation and access. Innovation policies stress that technology should be seen as an equalizer, not as an economic privilege. Avgerou (2010) suggests that government policies should include the innovation ecosystems of underserved communities. This can mean financing community-oriented data projects, providing grants to local entrepreneurs, and financing initiatives with direct social benefits. As part of the "Space for Inclusion" framework, both national space policies and poverty reduction strategies should be integrated (Brunswick, 2025).

Targeted subsidies, concessional loans, and blended finance create mechanisms that lower the cost of entry for economically marginalized areas. Universal satellite broadband requires the rural electrification analogy, meaning that public financing should create a low enough cost barrier to encourage private investment (Maîtrot et al., 2021). Tax breaks and innovation grants can encourage community-oriented space entrepreneurs to address their communities' social problems. The SERVIR initiative, a collaboration of USAID and NASA, provides satellite data for agriculture, water, and disaster management for regions in Africa, Asia, and Latin America (Becker-Reshef et al., 2010). In Kenya and Nepal, SERVIR has worked with smallholder farmers to enhance forecasting yields and drought resilience. Similarly, the Global Agricultural Monitoring (GEOGLAM) project uses EO and field data to support food security through early warnings (Rembold et al., 2019). The African Telehealth Network uses communication satellites to connect rural clinics and urban hospitals for remote diagnosis and consultations (WHO, 2025). In India, ISRO's Telemedicine Program connects over 100 hospitals via satellite links to provide remote specialized healthcare (Maral & Bousquet, 2020). UNESCO's Education for All via Satellite initiative began educational broadcasting to remote rural schools and teacher-training centers in Asia and Africa (UNESCO, 2023). During the COVID-19 pandemic, remote rural schools also used these satellite educational services to support learning where there was no terrestrial internet access. As Maîtrot et al. (2021) note, these services illustrate the equity dividend of satellite education: no learner is left behind, regardless of location.

The next frontier of inclusive development lies at the intersection of space-based systems and artificial intelligence (AI), machine learning, and big data analytics. Kim et al. (2022) illustrate how, with the use of AI satellite imaging, poverty mapping can be done at previously unattainable levels. Furthermore, these techniques can be used to predict and analyze changes in socioeconomic factors, pinpoint areas of vulnerability, and provide precise guidance for targeted interventions. Mulhaney et al. (2024) argue that the combination of AI and cloud computing has the potential to expand satellite technologies from local pilots to full nationwide programs. For example, predictive models using spatial-temporal data can forecast harvests, predict health issues, and even forecast migration. Such systems enable governments to deliver aid in a proactive, anticipatory manner, rather than simply reacting to issues, and thus, fit within the framework of resilience-based development (Hallegatte et al., 2020). Yet, future pathways need to address the preservation of bias and exclusion. As Heeks (2008) and Toyama (2015) note, the reliance on incomplete or biased data results in algorithmic bias and reinforces the status quo of inequality. To meet the standards for inclusive AI, communities must be actively involved in data governance, models must be transparent, and there must be ongoing ethical assessments. Brunswick (2025) proposes integrating technology guided by inclusivity, privacy, and sustainability, and therefore speaks to the need for responsible innovation ecosystems.

The way an institution approaches a problem can span a wide spectrum of methods and levels at which it uses technology. Approaches can be more than simply employing hardware or software. Space-based technologies (SBT) systems, and the use of satellites in conjunction with remote sensing and communication systems, have great potential. The focus should be more on building networks and systems of collaboration, and on empowering people in local areas. This requires the promotion of participatory governance or policy making, and certainly the embedding of ethics in technology. From the above, it can be seen why Sachs (2015) so aptly stated that for something to be sustainable, it must be achieved through solidarity.

VII. CONCLUSION

The current chapter will illustrate the prospective transformative evaluation of the rudimentary application of space-derived technology in alleviating poverty among the working poor and the underdeveloped. Remote sensing, earth observation, communication satellites, and geospatial mapping all exemplify formerly inaccessible and elitist technologies that have now become enablers of the empowerment of the marginalized. The chapter illustrates the ability of space-based systems to inform policy, improve and sustain livelihoods, and supply missing or underdeveloped services to the marginalized. The discussion demonstrated that combining such systems in the poverty and non-poverty paradox will increase agricultural productivity, improve resilience to climate and market shocks, enhance access to education, health, and financial services, and, most importantly, improve overall human dignity. Most importantly, the current chapter emphasized that most of the value of these space-based systems and tools derives from their social application and from the ability of such technologies to bridge the gap in dignity, opportunity, and information for all, including the marginalized.

Most importantly, the current chapter highlighted that most of the value of these systems and tools derives from their social application, enabling the marginalized to bridge gaps in dignity, opportunity, and information. The possible paths for space-enabled development greatly depend on how effectively the space-based development innovations are harmonized with the global social justice agenda to promote equity. This requires the global community not to focus on the advancement of technology for its own sake, but on innovations that promote inclusive growth and social justice. As the global community intensifies its engagement in the realms of the digital and space age, it should rightly prioritize the accessibility of data, the availability of digital and space-age tools, and the provision of digital literacy as public goods. It should morally empower the working poor by advocating and morally advocating for structures of inclusion, ethical governance, and strong capacity development. This development phase should promote the launch of more satellites and the creation of space-derived data value chains that empower local action by informing, teaching, guiding, and strengthening the role of community-based workers. It should help them to help the communities themselves.

This space-derived data value chain will help to close the gaps in spatial and digital equity. It will help people who are still grappling with the struggles of their daily lives. It will help to close the gaps in spatial and digital equity. The new challenge is for policymakers, technologists, and communities to address the existing gaps between the ability to do something and its accessibility. In this regard, policymakers must focus on a dual approach to inclusive innovation that brings together space-related technologies and development and situates innovation within relevant national development frameworks.

On the other hand, technologists must build systems that are inexpensive, adaptable, self-contained, and human-centered. To this end, the working poor must not be passive aid recipients, but respondents, so they can actively participate in shaping their own better futures. Furthermore, local communities and institutions need to be empowered so that they can own the technology and adapt, reinterpret, and integrate it within their cultural and economic systems. This integrated partnership is based on the collaboration that transcends disciplines and geography. It challenges the perception of technology as just progress and redefines it as a partnership, a collective responsibility to ensure that progress diminishes gaps rather than widens them.

This chapter seeks to contribute to the burgeoning dialogue on technology and social justice by reframing space technology from a high-tech, exclusive perspective to an inclusive, people-centric development solution. In this regard, it attempts to position satellites not as symbols of scientific prestige; rather, they are tools of human advancement that can shed light on the hidden geographies of poverty, civilize the disconnected, and galvanize people-centered, as opposed to productivity-centered, value systems. Therefore, this chapter argues that a defining characteristic of sustainable development for the twenty-first century will be driven not only by humanity's ability to reach and explore outer space, but also by the moral imperative to use that space to address humanity's most pressing challenge: leaving no one on the ground behind.

REFERENCES

Aggarwal, P., Vyas, S., Thornton, P., Campbell, B. M., & Kropff, M. (2019). Importance of considering technology growth in impact assessments of climate change on agriculture. *Global Food Security*, *23*, 41–48. DOI: 10.1016/j.gfs.2019.04.002

Avgerou, C. (2010). Discourses on ICT and development. *Information Technologies and International Development*, *6*(3), 1–23. https://eprints.lse.ac.uk/35564/

Banerjee, A., & Duflo, E. (2013). Poor economics. A radical rethinking of the way to fight global poverty'. *Society and Economy*, *35*(4), 573–587. https://www.ceeol.com/search/article-detail?id=622820

Becker-Reshef, I., Justice, C., Sullivan, M., Vermote, E., Tucker, C., Anyamba, A., Small, J., Pak, E., Masuoka, E., Schmaltz, J., Hansen, M., Pittman, K., Birkett, C., Williams, D., Reynolds, C., & Doorn, B. (2010). Monitoring global croplands with coarse resolution earth observations: The global agriculture monitoring (GLAM) project. *Remote Sensing (Basel)*, *2*(6), 1589–1609. DOI: 10.3390/rs2061589

Brickell, K., Picchioni, F., Natarajan, N., Guermond, V., Parsons, L., Zanello, G., & Bateman, M. (2020). Compounding crises of social reproduction: Microfinance, over-indebtedness and the COVID-19 pandemic. *World Development*, *136*, 1–4. DOI: 10.1016/j.worlddev.2020.105087 PMID: 32834384

Brunswick, S. (2025). *What's space got to do with it?: An interstellar guide to success.* SB Global LLC.

Brynjolfsson, E., & McAfee, A. (2014). *The second machine age: Work, progress, and prosperity in a time of brilliant technologies.* WW Norton & Company.

Bureau of Labor Statistics. (2024). *Industry and occupational employment projections overview and highlights, 2023–33.* U.S. Bureau of Labor Statistics. https://fraser.stlouisfed.org/files/docs/publications/bls_mlr/bls_mlr_20241107.pdf

Castells, M. (2010). *The rise of the network society* (2nd ed.). John Wiley & Sons.

Chen, R. (2019). *Policy and regulatory issues with digital businesses.* World Bank Policy Research Working Paper. https://lawscopelibrary.com/admin/articles/policy-and-regulatory-issues-with-digital-businesses/WPS8948.pdf

Dare, P. M., & Fraser, C. S. (2001). Cover: Mapping informal settlements using high resolution satellite imagery. *International Journal of Remote Sensing*, *22*(8), 1399–1401. DOI: 10.1080/01431160120654

Dercon, S. (2022). *Gambling on development. Why some countries win and others lose*. Hurst & Company.

Devereux, S., & Sabates-Wheeler, R. (2004). *Transformative social protection.* Institute of Development Studies. https://hdl.handle.net/20.500.12413/4071

FAO. (2022). *The state of food and agriculture 2022: Leveraging automation in agriculture for transforming agrifood systems*. Food and Agriculture Organization., DOI: 10.4060/cb9479en

Fields, G. S. (2012). *Working hard, working poor: A global journey*. Oxford University Press.

Friedman, T. L. (2006). *The world is flat: A brief history of the twenty-first century*. Farrar, Straus and Giroux.

Greene, M. (2018). Socio-technical transitions and dynamics in everyday consumption practice. *Global Environmental Change*, *52*, 1–9. DOI: 10.1016/j.gloenvcha.2018.05.007

Hallegatte, S., Vogt-Schilb, A., Rozenberg, J., Bangalore, M., & Beaudet, C. (2020). From poverty to disaster and back: A review of the literature. *Economics of Disasters and Climate Change*, *4*(1), 223–247. DOI: 10.1007/s41885-020-00060-5

Hargittai, E. (2003). The digital divide and what to do about it. In Jones, D. C. (Ed.), *New economy handbook* (Vol. 2003, pp. 821–839). Academic Press.

Heeks, R. (2008). ICT4D 2.0: The next phase of applying ICT for international development. *Computer, 41*(6), 26–33. DOI: 10.1109/MC.2008.192

Helsper, E. (2021). *The digital disconnect: The social causes and consequences of digital inequalities*. SAGE Publications. DOI: 10.4135/9781526492982

Henderson, J. V., Storeygard, A., & Weil, D. N. (2012). Measuring economic growth from outer space. *The American Economic Review, 102*(2), 994–1028. DOI: 10.1257/aer.102.2.994 PMID: 25067841

Hickel, J. (2017). *The divide: A brief guide to global inequality and its solutions*. Random House.

ILO. (2022). *World employment and social outlook: Trends 2022*. International Labour Organization., DOI: 10.54394/DSPL5113

ITU. (2023). *Measuring digital development: Facts and figures 2023*. International Telecommunication Union. https://www.itu.int/hub/publication/d-ind-ict_mdd-2023-1/

Jean, N., Burke, M., Xie, M., Davis, W. M. A., Lobell, D. B., & Ermon, S. (2016). Combining satellite imagery and machine learning to predict poverty. *Science, 353*(6301), 790–794. DOI: 10.1126/science.aaf7894 PMID: 27540167

Johnson-Freese, J. (2016). *Space warfare in the 21st century: Arming the heavens*. Routledge., DOI: 10.4324/9781315529172

Kabeer, N. (2015). Gender, poverty, and inequality: A brief history of feminist contributions in the field of international development. *Gender and Development, 23*(2), 189–205. DOI: 10.1080/13552074.2015.1062300

Kanbur, R., & Venables, A. J. (2005). Spatial inequality and development. In Kanbur, R., & Venables, A. J. (Eds.), *Spatial inequality and development* (pp. 1–17). OUP Oxford. DOI: 10.1093/0199278636.003.0001

Kerle, N. (2013). Remote sensing based post-disaster damage mapping with collaborative methods. In Zlatanova, S., Peters, R., Dilo, A., & Scholten, H. (Eds.), *Intelligent systems for crisis management: Geo-information for disaster management (Gi4DM) 2012* (pp. 121–133). Springer., DOI: 10.1007/978-3-642-33218-0_9

Kim, Y., Safikhani, A., & Tepe, E. (2022). Machine learning application to spatio-temporal modeling of urban growth. *Computers, Environment and Urban Systems, 94*, 1–12. DOI: 10.1016/j.compenvurbsys.2022.101801

Krige, J. (2019). *How knowledge moves: Writing the transnational history of science and technology*. University of Chicago Press. DOI: 10.7208/chicago/9780226606040.001.0001

Lund, F., & Nicholson, J. (2003). *Chains of production, ladders of protection: Social protection for workers in the informal economy*. School of Development Studies.

Maîtrot, M., Wood, G., & Devine, J. (2021). Understanding resilience: Lessons from lived experiences of extreme poverty in Bangladesh. *Development Policy Review, 39*(6), 894–910. DOI: 10.1111/dpr.12543

Maral, G., & Bousquet, M. (2020). *Satellite communications systems-Systems, techniques and technology* (6th ed.). Wiley. DOI: 10.1002/9781119673811

McGee, R., & Edwards, D. (2016). Introduction: Opening governance: Change, continuity and conceptual ambiguity. In D. Edwards & R. McGee (Eds.), *Opening governance* (Vol. 47, pp. 1–22). IDS Bulletin DOI: 10.19088/1968-2016.103

Mellor, J. W. (2017). *Agricultural development and economic transformation: Promoting growth with poverty reduction*. Palgrave Macmillan. DOI: 10.1007/978-3-319-65259-7

Meroni, M., Rembold, F., Urbano, F., Csak, G., Lemoine, G., Kerdiles, H., & Perez-Hoyos, A. (2019). *The warning classification scheme of ASAP–anomaly hot spots of agricultural production, v8. 0*. Luxembourg: Publications Office of the European Union. https://data.europa.eu/doi/10.2760/6814458

Milanovic, B. (2019). *Capitalism, alone: The future of the system that rules the world.* Harvard University Press.

Mulhaney, A., Parkinson, M., & Abdulfatah, A. K. (2024). A Human-Centric Approach to Digital Transformation. In A. Visvizi, O. Troisi, V. Corvello, & M. Grimaldi (Eds.), *Research and Innovation Forum 2024* (pp. 621–634). Springer. DOI: 10.1007/978-3-031-78623-5_49

Phillips, N. (2017). Power and inequality in the global political economy. *International Affairs, 93*(2), 429–444. DOI: 10.1093/ia/iix019

Ravallion, M. (2016). *The economics of poverty: History, measurement, and policy.* Oxford University Press. DOI: 10.1093/acprof:oso/9780190212766.001.0001

Rembold, F., Meroni, M., Urbano, F., Csak, G., Kerdiles, H., Perez-Hoyos, A., Lemoine, G., Leo, O., & Negre, T. (2019). ASAP: A new global early warning system to detect anomaly hot spots of agricultural production for food security analysis. *Agricultural Systems, 168*, 247–257. DOI: 10.1016/j.agsy.2018.07.002 PMID: 30774185

Rodgers, G. (2018). Inequality in the Indian growth regime. *Indian Journal of Human Development, 12*(2), 134–148. DOI: 10.1177/0973703018793956

Rostow, W. W. (1960). *The stages of economic growth: A non-communist manifesto.* Cambridge University Press.

Rostow, W. W. (2000). The stages of economic growth. In Sanderson, S. K. (Ed.), *Sociological worlds: Comparative and historical readings on society* (pp. 130–134). Routledge., DOI: 10.4324/9781315063362-14

Sachs, J. D. (2015). *The age of sustainable development.* Columbia University Press. DOI: 10.7312/sach17314

Sandau, R. (2010). Status and trends of small satellite missions for Earth observation. *acta astronautica, 66*(1-2), 1-12. DOI: 10.1016/j.actaastro.2009.06.008

Sayed, Z. (2015). Transforming our world: The 2030 agenda for sustainable development. *Ethics & Critical Thinking Journal, 2015*(3), 112–142.

Schwab, K. (2016). *The Fourth Industrial Revolution*. World Economic Forum.

Sen, A. (2015). Development as freedom (1999). In Roberts, J. T., Hite, A. B., & Chorev, N. (Eds.), *The globalization and development reader: Perspectives on development and global change* (2nd ed., pp. 525–548). Wiley Blackwell.

Sumner, A. (2016). *Global poverty: Deprivation, distribution, and development since the cold War*. Oxford University Press. DOI: 10.1093/acprof:oso/9780198703525.001.0001

Tennin, K. L., & Brunswick, S. (2024). Sustainable approaches to poverty and challenges from the 21st century. In Goel, R., Singh, T., Rahman, M. M., Islam, Q. T., & Baral, S. K. (Eds.), *Understanding the multidimensional nature of poverty* (pp. 13–41). Emerald Publishing., DOI: 10.1108/978-1-83753-292-620241002

Thornton, P. K., Ericksen, P. J., Herrero, M., & Challinor, A. J. (2014). Climate variability and vulnerability to climate change: A review. *Global Change Biology*, *20*(11), 3313–3328. DOI: 10.1111/gcb.12581 PMID: 24668802

Toyama, K. (2015). *Geek heresy: Rescuing social change from the cult of technology*. PublicAffairs.

UNDRR. (2023). *Global assessment report on disaster risk reduction 2023: Mapping resilience for the sustainable development goals*. Geneva: United Nations Office for Disaster Risk Reduction. https://digitallibrary.un.org/record/4022485?ln=en&v=pdf

UNESCO. (2023). *Global education monitoring report 2023: Technology in education: A tool on whose terms?* UNESCO. https://www.unesco.org/gem-report/en/technology

United Nations. (2015). *Transforming our world: The 2030 agenda for sustainable development*. United Nations. https://sdgs.un.org/2030agenda

UNOOSA. (2024). *From strategy to action: Annual report 2024*. Vienna: United Nations Office for Outer Space Affairs. https://www.unoosa.org/documents/pdf/annualreport/UNOOSA_Annual_Report_2024.pdf

U.S. Census Bureau. (2023). *Census bureau releases new report about persistent poverty at county and census-tract level.* United States Census Bureau. https://www.census.gov/newsroom/press-releases/2023/persistent-poverty.html

Warschauer, M. (2003). *Technology and social inclusion: Rethinking the digital divide.* MIT Press. DOI: 10.7551/mitpress/6699.001.0001

WHO. (2025). *Global strategy on digital health 2020-2027.* Geneva: World Health Organization. https://iris.who.int/server/api/core/bitstreams/a64d584f-015b-471f-9d86-9ea263cfb516/content

Worl Bank. (2022). *Poverty and shared prosperity 2022: Correcting course.* Washington DC: World Bank. https://openknowledge.worldbank.org/server/api/core/bitstreams/b96b361a-a806-5567-8e8a-b14392e11fa0/content

ADDITIONAL READING

Bhaskaranarayana, A., Varadarajan, C., & Hegde, V. (2009). Space-based societal applications—Relevance in developing countries. *acta astronautica, 65*(9–10), 1479–1486. DOI: 10.1016/j.actaastro.2009.03.080

Hargreaves, P. K., & Watmough, G. R. (2020). Satellite Earth observation of socioeconomic conditions for improved poverty reporting. *Proceedings: SPIE Space. Satellites and Sustainability, 11527*, 8–15. DOI: 10.1117/12.2570831

Watmough, G., & Marcinko, C. L. (2026). EO for poverty: Developing metrics to support decision making using earth observation. *Elsevier: Comprehensive Remote Sensing, 9*, 304–325. DOI: 10.1016/B978-0-443-13220-9.00033-0

KEY TERMS AND DEFINITIONS

Geospatial Data: This data describes information relevant to points on Earth's surface, collected via satellites, remote sensing, GPS, and mapping. It helps analyze spatial patterns, track changes, and aids decision-making in urbanization, agriculture, and disaster management.

Poverty Alleviation: Poverty alleviation involves policies, programs, and interventions to reduce poverty and enhance quality of life for impoverished populations. It supports income, job creation, healthcare, education, and technology access. The aim is to help individuals and communities achieve long-term economic independence and social integration.

Satellite Communication: This technology uses satellites orbiting Earth to send and receive signals for long-distance communication. It is especially useful in rural areas where standard communication is lacking. Satellite communication supports internet, telecommunications, broadcasting, telemedicine, and emergency services.

Space-Based Technologies: An example is the ability to analyze Earth from space and track objects to assist in global communication. Space-based technologies are useful in climate monitoring and disaster management as they provide tools to observe and analyze events on Earth. These technologies are helpful for policymakers and researchers alike in developing solutions to climate change, managing disasters, and improving communications.

Sustainable Development: Sustainable development meets present economic, social, and environmental needs without harming future prospects, emphasizing equity between now and the future. It involves growth aligned with social equity, environmental protection, and resource stewardship. The UN SDGs guide this approach.

Working Poor: The working poor are full- and part-time workers below the poverty line due to low pay, unstable employment, and lack of social protection. They face issues similar to those in the informal economy: inadequate social safety nets and low income, making it hard to meet basic living standards.

Chapter 3
Geospatial Mapping of Working Poverty and Socioeconomic Inequality

Ayesha Farooqi
http://orcid.org/0009-0000-4556-4211
Poornima University, Jaipur, India

Nikhil Kumar Goyal
http://orcid.org/0009-0007-4532-8033
Noida International University, Greater Noida, India

ABSTRACT

In this chapter, the geospatial technologies are discussed in terms of understanding and analysis of working poverty and socioeconomic inequality. Indeed, the spatial distribution of poverty and vulnerable populations can be visualized on the city and rural terrain by combining satellite imagery, remote sensing, and Geographic Information Systems (GIS) in order to locate vulnerable groups in both urban and rural environments. The chapter discusses relevant socioeconomic engines, which include the dynamic labor market, education, and access to infrastructure, and some of these spaces depict areas of inequality. Moreover, it gives the implications towards policymakers and planners set in evidence-based

DOI: 10.4018/979-8-3373-7118-4.ch003

interventions and focused programs. Emerging technologies, including machine learning and mobile data integration, are also singled out as to be used to monitor poverty in real time, with a feeling of providing a means of making use of more inclusive and useful approaches to reduce poverty.

1. INTRODUCTION TO WORKING POVERTY AND SOCIOECONOMIC INEQUALITY

1.1 Understanding Working Poverty

Working poverty is a situation in which people are in the employment but are unable to earn enough money to live at a poor standard. The working poverty, unlike the absolute poverty, is concerned with how the level of wages and employment provisions of an individual can make him or her gain access to economic security in terms of the minimum subsistence levels. The globe has millions of employees falling in such category particularly in the informal sector, low wages and precarious work systems. Working poverty has come to be a severe socioeconomic challenge and it crowds inadequacies in the labor markets, pay and social security.

The economic stability and social development have significant implications that are associated with this phenomenon. Low-income employees who are involved in low-paid jobs lack access to medical and educational services, shelter and become trapped in the circles of poverty over the generations. Further, the working poverty in question is not similar in terms of geographical territories; it is often concentrated in some areas, in cities slums or villages where individuals have no economical perspectives. These patterns should be interpreted in order to draft particular interventions, as well as determine the success or failure in the realization of the inclusive economic growth (Tokman, 1991).

It is significant, research-wise, that one should study the phenomenon of working poverty interdisciplinarily. Economics, sociology and public policy can be used in understanding the dynamics of income and employment and systems of welfare. However, the standard methods that employ surveys would omit the spatial dimension of the practice, and it may be hard to prove the localized differences. This weakness defines the significance of geospatial techniques, which offer the analysis ability of mapping poverty in space not just to the determination of who is poor but also where the poverty is more centralized (Floro, 1995).

1.2 Socioeconomic Inequality in Dimensions

Not only does income difference constitute socioeconomic inequality, but education, health, social capital, and access to basic services are all variations of socioeconomic inequality. It is shown in the disproportional opportunities, limited mobility, and unequal distributions of the assets, which in the majority of instances is intertwined with the other factors such as gender, ethnicity and geographical location. The sub-range of working poverty under further subdivision of this elevated inequality puts emphasis on the fact that the populations of the hired to work but marginalized in the labor markets workers are more exposed in comparison to other category of people.

Figure 1. Dimensions of inequality

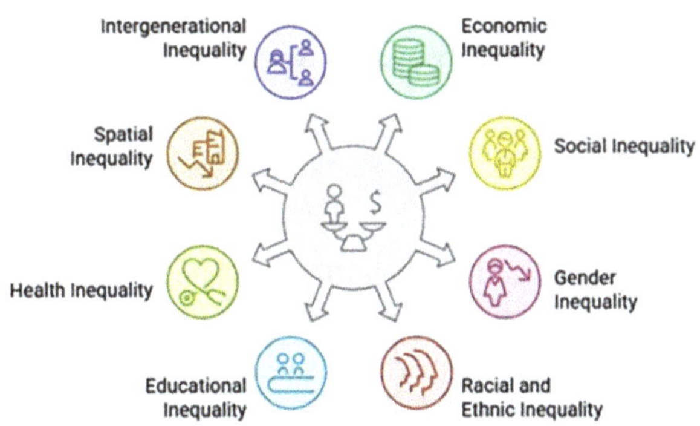

The issue of inequality can be considered on more than just a spatial level. On the domestic level it is manifested in domestic consumption, wealth accumulation and accessibility to basic services. On the local or regional level, it is manifested through a skewed infrastructure, employment and provision of services. Geospatial analysis has afforded the researcher with the capability to quantify these inequalities and the effects on spatial factors such as nearness to economic centres, availability of transportation, or like the population density on the socioeconomic results. This makes the policymakers get a better picture of inequality due to geographical factors by relating the two back to poverty (International Labour Organization, 1998).

In addition to that, inequality is dynamic and is developed under the influence of economic and political and social changes. Inequality Globalization, urbanization, and change in technology can also contribute to disparity by producing unequal access to high-value employment and other necessities. The spatial aspect toward inequality consequently becomes a key aspect to comprehend when attempting to formulate, responsive, and context-responsive policies.

1.3 The Geospatial Analysis Relevance

Geospatial analysis is a branch of analysis that takes advantage of Geographic Information Systems (GIS), satellite imagery, and remote sensing in mapping and analyzing the distribution of the social and economic variables in space. In the case of the working poverty, in this way, researchers can find out geographical hotspots, quantify distance to services, and spatial relationships between poverty and infrastructure or employment opportunities. However, unlike traditional survey tools, geospatial tools may be used to cover extensive reserves, make use of real-time data, and visualize trends that do not easily emerge in tabular data.

The strength of geospatial techniques is that they can be able to combine various layers of information. Using census, satellite-acquired nightlight intensity and maps of the transportation network as an example, a detailed map of working poverty concentrated by infrastructure and urban development can be created. Such a multidimensional view is of immeasurable value to policymakers, urban planners and social development agencies aiming to develop specific interventions and to spread resources effectively (Charmes, 1999).

Moreover, geospatial analysis promotes monitoring and evaluation since it allows tracking the change in poverty over time. The stakeholders may evaluate the competence of the programs and emphasize the strategies to aid the program implementation by monitoring the spatial shifts in response to policy interventions, economic shifts, or urban growth. The visualization of these patterns also improves the communication with the non-technical audience and makes the difference between understanding the data and taking the action.

2. GEOSPATIAL DATA SOURCES AND METHODOLOGIES

2.1 Satellite and Remote Sensing Data

The use of satellite imagery, remote sensing technologies has become essential in the process of mapping working poverty and the social economic inequality. The technologies measure a broad spectrum of environmental and human-related measures of economic activity, distribution, and development of infrastructure. Findings of light intensity at night, e.g. can also be a proxy of economic activity, brighter neighborhoods suggest more economic activity and darker areas can be indicators of poverty or under-development. Equally to this, land use and land cover maps created using satellite data may be used to determine informal settlements as well as patterns of agriculture and urban sprawl, which have a close connection with economic vulnerability.

The remote sensing data is high-resolution and real-time information across a wide geographical region enabling scholars to research in situ where the traditional data gathering is challenging or uncommon. Through the study of the temporal variations in the satellite images, urbanization, or increase in infrastructure or environmental degradation is traceable; all these will impact on working poverty. Also, the multispectral and hyperspectral imagery may showcase small-scale socioeconomic trends, including variation in the quality of housing or access to services, that are not noticeable through traditional data sets. This renders remote sensing as a useful instrument in the capturing of the spatial and temporal aspects of poverty (Lopez & Valdes, 1997).

Subsequently, however, remote sensing information should also be calibrated and compared with information on the ground. In unverified cases, local conditions could be lost and especially in local densely populated and haphazard rural areas, or in informal settlements. A combination of field survey or census data and satellite data enhances the accuracy and makes geospatial analysis to be a reflection of socioeconomic realities on the ground (Kabeer, 2003).

2.2 Techniques of Geographic Information Systems (GIS)

The analysis of working poverty and social economic inequality has a spatial dimension, which is provided by Geographic Information Systems (GIS). GIS allows to combine, visualize and analyze various layers of data, including demographics statistics up to infrastructure networks, environmental conditions and economic indicators. By means of GIS, scientists can overlap data, carry out spatial queries, and determine the correlation between poverty and such variables as proximity to markets, schools, or healthcare centers.

Hotspots analysis, spatial autocorrelation, and the analysis by using the Kernel density estimation are the spatial analysis methods used in GIS to identify the presence of clusters of poverty and the locality of extreme levels of socioeconomic deprivation. An example is the hotspot analysis that can be used to identify the locations of low-wage employment to help make policy intervention at the places where it is concentrated. On the same note, network analysis can be used to study how accessibility to roadways, public transport or economic facilities affect the level of income derived and hence ways in which geographic barriers to economic mobility exist (Duflo, 2012).

Scenario modeling is also facilitated by GIS which enables researchers and policymakers to model the impacts of the infrastructure development on employment programs or social policy on poverty patterns. GIS can be used as both a predictive and diagnostic tool, by visualizing the possible operations, which can be used in strategic planning and distribution of resources. Analytical rigor and visualization power render GIS important in converting intricate information regarding space to actionable information.

2.3 Data Integration and Validation

Geospatial analysis is powerful as it is able to combine various sources of data. With satellite imagery, census, household questions, administrative data, and crowd-sourced information should be merged to create a complete spatial implication of working poverty. By combining heterogeneous data, it is possible to achieve more precise mapping, since all sources partially address the shortcomings of the other ones. As an illustration, census data give information about the demographics and income but satellite data give the location distribution of settlement and infrastructure and mobile phone information can give the real-time data on economic life or population mobilization (International Labour Organization, 2012).

Validation is one of the important factors of reliability of integrated geospatial datasets. Field surveys and ground-truthing methods and comparison with official statistics assist in assuring that the patterns revealed in satellite and GIS analyses are a good representation of socioeconomic reality. Machine learning algorithms may be used to clean and harmonize large datasets, identify anomalies, and improve the predictive accuracy also. Sound validation will make sure the policymakers and planners will depend on these analyses to make sound decisions instead of basing interventions due to the possible misleading and incomplete information.

Additionally, the collection and incorporation of the spatial data should be ethically oriented towards such issues as the privacy of the data and protection of the vulnerable groups. Anonymity and bottlenecking of sensitive information is a must especially when dealing with granular datasets that in turn may identify socioeconomic circumstances in the household or individual level. Strict validation and ethical protection will contribute to the reliability and relevance of the geospatial poverty mapping in the society (United Nations Economic and Social Commission for Western Asia, 2014).

3. MAPPING AND VISUALIZING POVERTY PATTERNS

3.1 Spatial Distribution of Income and Employment

The functional knowledge of the spatial distribution of income and employment is the key to determining the areas of working poverty. Geospatial analysis allows researchers to provide images of differences between regions, neighborhoods, and even single-family households. With the help of mapping the income levels and employment density, one will be able to observe where economic opportunities are delivered and where they are not. As an example, employment rates and income diversity tend to be greater in urban centers, whereas agglomerates of low-income employees and precarious jobs may be observed in the peripheral urban centers or rural areas.

When census data or house holds survey is used to provide income distribution maps and then overlaid with the employment statistics, trends otherwise obscure by the traditional tabular data are revealed. High Employment and Low wages areas reflect high levels of working poverty in regions, which imply structural labour market problems. Conversely, the areas where the employment rate and the income level is poor indicate a general socioeconomic vulnerability which in majority of the instances, requires interventions in a multidimensional nature. As these spatial differences are captured, the policymakers will be able to place resources first and to develop certain programs to address the special needs of various communities (International Labour Organization, 2018).

Moreover, with the help of GIS, time analysis is possible, therefore, geospatial data might be utilized to trace the dynamics in relation to the rates of employment, as well as income distribution with time. It can especially be useful in establishing the impacts of economic policy, development projects or infrastructure projects. On the one hand, dynamic visualizations can demonstrate whether new industrial areas, transport connections, or social projects are successfully decreasing working poverty, or inequalities are still present. This type of insights will be essential in evidence-based planning and strategic intervention.

3.2 Poverty in Cities and Rural Areas

The poverty is faced differently in urban setting as compared to rural setting and this is because of the density of population, access to infrastructure, the labor markets and the social services. City poverty tends to be clustered within informal settlements, slums, and low-income neighborhoods that are expensive to live in and have limited opportunities of formal employment. There is usually poor sanitation, health, and education infrastructure facilities in these areas thereby increasing the socioeconomic inequality even though they are close to economic centers.

In contrast, rural poverty is typified by poor accessibility to markets, low agricultural productivity as well as inadequate infrastructure. Geographic isolation has the potential to strengthen working poverty, because the people living in those areas pay more to travel to work, and have less access to job opportunities, as well as to financial institutions. Geospatial analysis makes it possible as researchers can map these urban-rural disparities, and such methods as spatial clustering, density mapping, and travel-time analysis can be used to identify areas of concentrated vulnerability (International Labour Organization, 2019).

The patterns of urban and rural poverty are also helpful in comparative analysis through visualization - this is because the policies and interventions required to be context-specific. An example is that microfinance efforts could be better in the countryside, and urban slum upgrading efforts could be used in slum informal settlements. Geospatial mapping allows consideration of the details of spatial change, providing the strategies to reduce poverty with the focus on local contexts instead of universal solutions (International Labour Organization, 2020).

3.3. Hotspot and Cluster Analysis of Poverty

Hotspot and cluster analysis are needed techniques of determining where working poverty is concentrated. Through the spatial statistics, the researchers are able to identify the important groups of low-income households, high informal employment regions, or in regions that lack basic facilities. Such analyses point out the areas in which interventions are most likely to be found to be most effective and efficient in resource allocation.

Indicatively, the use of a hotspot analysis could demonstrate that some local areas in a city have a high unemployment rate as well as poor infrastructure, thus, necessitating a combination of economic and social policies. The cluster analysis could also present latent patterns such as regions of relative affluence amid the poverty that would have other wise not been identified in an aggregate statistics. The lessons may be applicable in the creation of the multi-dimensional intervention, such like special skills improvement courses, the investment in infrastructure, or the local social services (World Bank, 2013).

Figure 2. Hotspot analysis of poverty rates in South Carolina

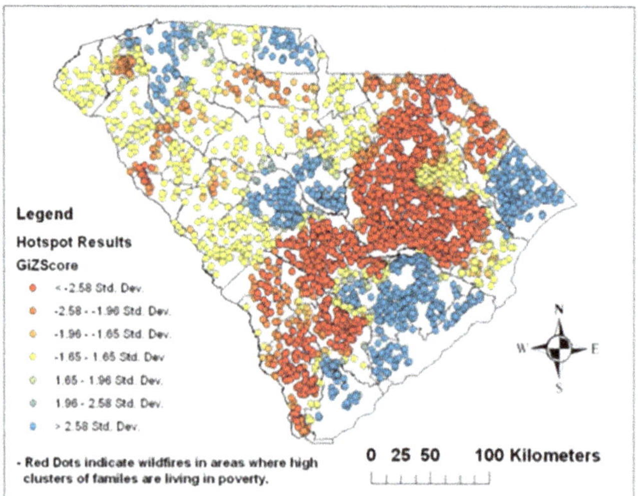

Furthermore, using hotspot analysis in combination with time information will enable policy makers to trace the changes in which poverty trends vary with time. It can demonstrate how effective the current interventions are and be utilized in predicting the new points of weaknesses. With the conceptualization of poverty as variable, spatially patterned, by visualizing poverty as a dynamic, spatially moving process, geospatial methods empower the action plan by transforming abstract socioeconomic representations into practicality in policy decisions, and in the process, the long-term strategic plan (World Bank, 2021).

3.4 Tools and Techniques of Visualization

The use of effective visualization is essential in conveying the different complex patterns of poverty to the policy makers, the researchers, and the public. Heat maps, choropleth maps, and interactive dashboards all result in the easy visualization of income inequalities, employee clusters, and service accessibility. Developed GIS systems also facilitate 3D modeling, network analysis, and simulation of scenarios and allow a deeper insight into the specifics of the spatial processes and possible interventions.

The interactive nature of visualizations improves decision-making by enabling the stakeholders to interact with data on various levels (country-wide to a neighborhood). They also assist in involving the stake holders and consequently the members of the community, the local government and the development agencies in identification of the areas of priorities. These tools also shorten the gap between geospatial research and implementation of policies because they transform incomprehensible forms of naked geospatial data into forms that can be interpreted and operated on (United Nations Development Programme, 2019).

In addition, the visualization tools can integrate predictive analytical and machine learning results providing insights into the future of swift destitution. This enables the proactive response in lieu of the reaction to minimise socioeconomic inequality, and subsequently measures of sustainable development that would anticipate and manage it beforehand. Such combination of geospatial visualization and predictive modeling will warrant that poverty mapping will not only be descriptive but prescriptive enough to offer both short run and long run action and planning.

4. SOCIOECONOMIC DRIVERS OF WORKING POVERTY

4.1 Labor Market Dynamics

The working poverty is constituted mainly by the labor market processes. These factors are the quality of employment, level of income, employment stability, and informal employment status which determines the capacity of individuals to be financially independent even in the employment process. Low incomes in the informal sector, in temporary contract or many more are prevalent in most regions that exposes the workers to alterations in their income and lack of social policies that would offer security such as health coverage or pension schemes. At the formal employment sectors, the forces that contribute to working poverty are wage stagnation and underemployment.

Geospatial analysis allows the researcher to examine the geo-distribution of workforce and the territory where precarious employment constitutes the majority. With a case in point, the satellite images employed in conjunction with the census information can be utilized to push forward the industrial regions where the employment rates are high, and the wages are low or those, in which informal labor predominates. Being informed about such trends in space assists policy makers to concentrate well on intervention, which can be stimulating the growth of skills programs within low labour neighbourhoods or stimulating the introduction of a high-quality job within economically deprived areas (United Nations Development Programme, 2020).

In addition, working poverty based on labor mobility, also depends on commuting. Inadequate transportation facilities or high transportation costs can also be a factor low access to higher remuneration and the employee can remain in the low-income neighborhoods. Geospatial techniques would help identify obstacles to economic vulnerability through mapping access to employment hubs. Such understanding can be used to design specific solutions to improve the situation, like upgrading the public transportation system or decentralizing labor market opportunity to diminish spatial inequalities (Organisation for Economic Co-operation and Development, 2019).

4.2 Distribution of Education and Skills

One of the major determinants of an individual in escaping working poverty is education and skills. The absence of quality education or job training usually traps the workers into low wage positions, continuing income incompetency loops. Socioeconomic inequality in much of the areas takes the forms of the imbalanced distribution of educational facilities where the rural areas and marginalized urban populations have little educational opportunities either through access to schools, training facilities either technical or higher educational facilities.

Education and working poverty may be shown to be correlated using geospatial mapping. An example can be provided using the overlay maps showing the distribution of schools, literacy rates and employment levels to demonstrate areas where low level of education is a source of economic vulnerability. This enables governments and other developmental agencies to determine the underutilized locations and invest heavily on educational facilities, skills development course, and adult education programs. Also, the appreciation of spatial patterns in education may be utilized in policy making on the equitable distribution of resources so that interventions are brought to the populations that are more at risk of staying in low-wage work (Tokman, 1989).

Economic outcomes are also influenced in the provision of skills within the workforce. Low levels of human capital in certain areas also tend to be incapable of attracting high level industries and supporting entrepreneurial activities, which reinforce nests of spatial poverty. The ability to map the level of skills and determine economic indicators allows developing the workforce specifically, so that training efforts match the available employment opportunities and the goals of the local development.

4.3 Access to Infrastructure and Public Services

Working poverty is highly affected by access to infrastructure and other amenities. Basic facilities like transport, health, electricity, water, and integration of financial services have an impact on productivity, income generation, and the quality of life in general. The unavailability of such resources can lead to higher living expenses, lower chances of employment, and heighten the susceptibility of the low-income employees. An example of this is that a rural region with poor market or road access would mean limited or constrained residents to low-income agricultural work, or an urban neighborhood with lack of healthcare or childcare facilities would limit the workforce, particularly among women.

The power of Geospatial analysis lies in the evaluation of the inequality of infrastructure. Using maps of transportation systems, healthcare facilities, and coverage of utilities, the researchers have the possibility to compare them with poverty data to determine the areas where poor service delivery leads to economic vulnerability. Spatial accessibility indicators (travel time to the closest hospital or work center) inform about structural inequalities promoting working poverty. The insights can be used to conduct specific interventions, e.g. the focus on development of infrastructure, development of the public services or social protection in under-served regions (Fields, 2004).

Besides, infrastructure and services are interrelated with other socioeconomic drivers. As an example, the educational facilities can be enhanced, but until access to transport is enhanced, the opposite side of the coin, expanding working opportunities cannot be effective in the absence of affordable electricity or online connectivity. By geospatially mapping these interdependences a comprehensive approach is possible to working poverty and the integrated policy responses will take into account working poverty across many different dimensions.

4.4 Socioeconomic Shocks and vulnerability

Exposure to socio economic shocks such as economic recessions, natural disasters, health crises and market volatilities determine working poverty as well. The people who have low income, savings, or social protection are also susceptible to these shocks because one negative event can cause them to further descend into economic insecurity. As an example, seasonal jobs in the agricultural sector or other informal sectors can be very vulnerable to weather conditions, commodity prices or world supply chain breakages.

Geospatial methods may be able to find the regions that are vulnerable in socioeconomic terms, as well as those that are susceptible to shocks. The combination of mapping historical disasters, climate risks, or economic crises and poverty indicators will enable policymakers to see the populations under the most threat. This will help to create special resilience plans, e.g. disaster preparedness, microfinance plans, or local social safety netting plans which will be able to cushion against immediate economic losses (Squire, 1996).

Geospatial analysis through the combination of these insights and information regarding labor markets, education, and infrastructure offers the multidimensional representation of working poverty. It does not just show the location of the vulnerability but also why it continues to exist, which leads to more specific and effective interventions which do not deal with the symptoms to be better, but rather with the source of the vulnerability.

5. POLICY IMPLICATIONS AND TARGETED INTERVENTIONS

5.1 Evidence-Based Planning

The best place to start effective poverty alleviation is at the level of evidence based planning which places emphasis on accurate, timely, and spatially detailed data to make a policy decision. Geospatial mapping presents a decisive benefit in shedding light on the geographical spread of working poverty and social economic disparities whereby policymakers are enabled to find areas of priority as well as to allocate resources in an efficient manner. Determining how the income distribution, patterns of employment and access to infrastructure are distributed enables decision-makers to abandon a one-size-fits-all approach in favor of a localized intervention that meets community-specific requirements.

To provide an example, a city-based analysis would show groups of poor families in the industrial centers but without means of transportation.

The planners can use this information to plan certain programs such as subsidized transport routes, vocational training programs based on the available employment opportunities. To the same degree, low-cost connected rural areas could be enhanced or with the supply of mobile services. The poverty reduction strategies are also evidence-based which ensures that in addition to being well targeted interventions are also also cost-effective and maximize the benefit of the limited resources available and the overall effectiveness of the strategies applied (Smith & Ward, 2014).

Moreover, it would be possible to utilize the geospatial information and adapt the temporal one to develop dynamic planning. The policymakers can keep track of the development of poverty over time, the success of the interventions, and take appropriate measures. This is to encourage resilience and responsiveness whereby the policies are developed in response to altered socioeconomic realities and emergent challenges.

5.2 Poverty Reduction Interventions

Elimination of working poverty requires specific programs to reduce poverty which must be critical to move the economy to greater economic mobility. These programs may include; jobs creation, skills building programs, social security programs, and access to basic needs. The geospatial analysis applies in the development of the most required areas which require such programs and even the monitoring of these programs.

The ideas of employment-based intervention, such as job training facilities or development of microenterprises can be strategically located in regions with large populations of low-wage workers or informals. Similarly, social protection programs, like conditional cash transfer or food aid can be allocated some priority to the worst hit communities in terms of the signs of poverty. The integrations of spatial data of income, employment and infrastructure help the policymakers to come up with multi-dimensional programs that are able to reduce not only the shortage of income in short run, but also also the structural factors that constitute the perpetuation of the poverty (Tatem et al., 2014).

In addition to that, there is flexibility in the control of the program with geospatial surveillance. As an example, satellite images and GIS mapping can determine changes in the settlement patterns, urbanization or accessibility of communal services that will be fed back on whether the interventions are accessing the target populations. These are evidence-based changes that would make sure that poverty alleviation activities are effective, fair, and differentiated to respond to the real-life scenarios.

5.3 Strategies of Monitoring and Evaluation

Well developed monitoring and evaluation (M&E) systems are essential in poverty reduction efforts to make them successful and sustainable. Geospatial methods improve on the M&E by offering both visual and quantitative, and spatially explicit indicators of the outcomes of the program. Using the example of the pre- and post-intervention maps of the employment distribution once the programs are implemented, the policymakers will be able to evaluate whether the programs are successful in meeting their goals and where more support will be necessary.

In addition, geospatial M&E allows identifying unanticipated effects or program inequality. Certain interventions can be biased to other areas and people, abandoning some. Spatial disaggregated information is useful in recognising these gaps where corrections can be made to the course and just redistribution of resources can be made. GIS dashboards as interactive systems that are linked with real-time information sources like surveys, mobile devices, or satellite pictures are the ones that allow monitoring on a continuous basis and ensure that policymakers can act immediately and appropriately in response to the evolving situations (Elbers et al., 2003).

In addition to program evaluation, geospatial M&E helps to promote accountability and transparency. The results of the the projects can also be easily available to the stakeholders consisting of local communities, civil society groups and government agencies because it is easy to visualize the results of the poverty alleviation efforts. Such transparency gives rise to trust and stakeholder involvement and greater policy intervention legitimacy. By a mixture of extensive spatial research along with comprehensive monitoring, the governments and development agencies will be in a position to verify that its effort is considerable on working poverty and to refer to socioeconomic equalities.

5.4 Integrated Policy Approaches

Poverty is multidimensional with the influence of the labor markets, education, infrastructure, and existence of the socioeconomic shocks. This implies that effective interventions can only be achieved when the compound policy interventions address different drivers simultaneously. This kind of integration is only possible with the assistance of geospatial analysis that determines the areas that can be impacted the most with the use of multiple layers of socioeconomic data.

One such example would be an initiative that aims at mediating skills training and transportation allowances and microfinance may target convergence of the low educational level, mobility, and access to credit in a region. Governments can build synergistic effects by creating policies that deal with issues that are not isolated to enhance both the short-term and the long-term resilience. Integrated strategies also contribute to policy fragmentation avoiding interventions that oppose each other (Barrington-Leigh & Millard-Ball, 2017).

Moreover, geospatial information can inform cross-sector cooperation of urban planners, social welfare organizations, and economic development organizations. Shared spatial datasets offer a mutual insight into how vulnerabilities occur and provides the ability to act efficiently as well as in a manner that is contextual. Geospatial evidence-based integrated policy efforts, therefore, are the best way forward towards working poverty alleviation and reduction of socioeconomic disparity.

6. EMERGING TRENDS AND FUTURE DIRECTIONS IN GEOSPATIAL POVERTY MAPPING

6.1 Machine Learning and Predictive Analytics

The research of working poverty is being transformed by machine learning (ML) and predictive analytics to allow researchers to derive actionable insights on large and complex datasets. Old-fashioned approaches might be based on the census data or a fixed survey which might be out-of-date or granularity might be inadequate. Conversely, ML algorithms are capable of using a wide range of different sources of data to forecast high-risk poverty locations based on satellite images, mobile phone activity, financial activities, and social media--which makes them highly accurate in their predictions.

Predictive models are able not only to concentrate on the existing poverty hotspots, but also those that may be subjected to the deteriorating conditions as a result of economic shocks, climate change, or urbanization patterns. As an illustration, the intensity of light at night and land use patterns based on nightlight as applied by the ML algorithms can predict economic vulnerability in neighborhoods, thus informing policy interventions in advance. In addition, predictive analytics can be used in scenario modelling to enable policymakers to model the effect of an infrastructure project or employment initiative or social safety net before they are put into action. This initiative will promote efficiency and fairness in the measures of poverty alleviation (Feenstra et al., 2015).

Nevertheless, the accuracy of ML in geospatial poverty mapping categorizes on the accuracy and heterogeneity of input data. The messages that are not collected without bias, the underrepresentation of marginalized groups of people, or the absence of information may provoke false forecasts. Integrating machine learning with the field surveys and validation should be considered a pillar of robustness and credibility so that the use of AI-based insights may be trusted by the regulators, and the human control is provided.

6.2 Crowdsourced and Mobile Data Integration

The data crowdsourced and mobile technology are becoming strong complements to the traditional geospatial data. Mobile devices, GPS, and local reporting tools can produce immense real-time information on spatial population movement, their activity, and access to services. Located, often inaccessible in official datasets and mostly in informal neighborhoods or rural regions, crowdsourced mapping projects, e.g. participatory mapping or community surveys, offer local information.

Specifics The granularity and up to date of poverty mapping is improved by integrating mobile and crowdsourced data. As the examples above can illustrate, mobility data can be used to understand commuter issues among low-wage workers, and participatory mapping can be used to recognize informal service deficiencies, such as unregistered clinics or unpaved roads. Such information can enable the planners to develop interventions that are not only evidence-based but also evidence-based grounded on the local realities so that their relevance and effectiveness would be higher (Alkire & Foster, 2011).

Also, crowdsourced data can bring communities together, as it allows officials to engage communities in the mapping project and gives them the opportunity to express their needs. These forms of participation enhance transparency, trust between the policymakers and people, as well as the establishment of feedback, which enhances development programs. Crowdsourced intelligence together with conventional geospatial data can support the work by the researcher to a deeper, more detailed sense of working poverty and socioeconomic inequality.

6.3 Moving Towards Inclusive and Real Time Poverty Mapping

The future of geospatial poverty mapping with integration of individual data source and methods of analysis lies in the development of inclusive real-time monitoring systems. Real time mapping can help governments, non-governmental organizations, and international agencies to immediately address emerging weaknesses, including the ones triggered by economic crises, natural disasters, or a health crisis among the population. Through constant modernization of spatial databases, a policy developer can monitor the changes caused by interventions, detect any gaps, or change approaches in real time.

Inclusivity in poverty mapping In poverty mapping, all groups in society such as the marginalized and inaccessible are also incorporated in collection and analysis of data. This is assisted by developments in remote sensing, crowdsourcing and mobile technologies as they facilitate in the capture of information in places that traditional surveys might miss. An example would be that informal settlements are not always visible within national data, but real-time geospatial analysis can be used to establish the geography, population density and infrastructural requirements which can be used to intervene with those most in need (Gibson et al., 2010).

Combining artificial intelligence, predictive analytics, and real-time data platforms is likely to suggest a future in which poverty mapping is a prescriptive in addition to a descriptive activity. The decision-makers will be in a position to foresee the trends of vulnerability, efficiently deploy the resources and also keep track of the results. Through a technological innovation coupled with participatory strategies, geospatial poverty mapping will be an active tool towards pursuing socioeconomic equity, resilience, and sustainable development objectives at local and global portfolios.

7. CHALLENGES AND LIMITATIONS IN GEOSPATIAL POVERTY MAPPING

7.1 Data Availability and Quality

Availability and quality of the data is one of the main issues in geospatial poverty mapping. Not all sectors have reliable socioeconomic and spatial data, specifically, low-income states or rural regions. Census information can be old-fashioned, partial or the coarse scale on which it is aggregated is too elevated to show significant local trends. Although the information obtained by satellites and remote sensing is rich, the data might be less detailed to identify the subtle socioeconomic circumstances, including informal housing standards or household income rates.

Poor quality of data is also caused by unequalities, errors, and the absence of values as they result in poor analysis or false interpretation. A combination of several datasets, including household surveys, administrative records, as well as satellite images can soothe these problems but the verification, calibration, and standardization of the process must be attentively validated. Policy advice based on geospatial studies can be useless or unfair, without careful consideration of the quality of this data (Ravallion, 2016).

7.2 Technical and Analytical Problems

Mapping of geospatial poverty is a complex process, which uses advanced computing techniques, such as GIS, remote sensing, machine learning, and predictive modeling. The following factors are analytical: model the dynamic socioeconomic processes, and controlling confounded variables, as well as the synthesis of heterogeneous datasets of divergent spatial and temporal resolutions (De Sherbinin, 2018).

Besides, geospatial outputs must be interpreted with caution. Spatial relations are not necessarily causal but tackle a correlation and visualizations simplify the objective of complicated things. As an example, high rate of low-income households in the city could be the results of past settlement development instead of existing economic status. To be able to draw valuable and accurate interpretation, analysts have to merge geospatial understanding and field validation with both field knowledge and theory of socioeconomic understanding (Jerven, 2013).

7.3 Socio-Political and Implementation Barriers

Lastly, geospatial poverty mapping can be constrained by the socio-political reasons. In parts of the world, politics, this could be political opposition, inability of institutions, or scarcity of resources could act as barriers to collecting, distributing and using geospatial data. Also, there are no clear-cut methods to translate spatial knowledge into action, local governance, and culture, and challenges in their administration may influence the design and implementation of interventions (Jean et al., 2016).

To overcome these barriers, it is vital to implement cross-sector collaboration, capacity-building, as well as the establishment of the institutional frameworks that could facilitate the making of the data-driven decisions. New technological solutions along with governance changes and involvement of stakeholders guarantee the fact that geospatial poverty mapping would be able to translate into meaningful changes benefiting vulnerable populations in the real world (Engstrom et al., 2017).

CONCLUSION

Geospatial mapping can provide a revolutionary perspective to learning and solving working poverty and socioeconomic inequality. Quantifying patterns of poverty, defining vulnerable populations, and creating specific interventions with the help of combined satellite imagery, remote sensing, GIS, and various socioeconomic data, researchers and policymakers will be able to visualize them. Geospatial analysis is important with its variations highlighted as big in this chapter, but has demonstrated that the technology is useful in making sense of the inequality in the urban and rural regions, in mapping the areas of income and employment, and plotting the locations of key infrastructures and services as far as their accessibility is concerned. Such types of spatial knowledge play an invaluable role in evidence-based planning processes, as well as resource allocation to the majority of the communities in need.

Accuracy and timeliness in relation to mapping poverty is enhanced due to the application of new technologies like machine learning, predictive analytics, and crowd sourced data. These new technologies can enable the live tracking, mapping the out in every scenario, predicting the socioeconomic shocks and give policy-makers with the chance to act proactively, taking context-specific actions. The local vulnerabilities can be addressed by having geospatial mapping of poverty, which results in the integration of technical and the participatory, and provides a viable actionable information that can influence sustainable and inclusive development.

Despite the opportunities, geospatial poverty mapping has had many challenges including quality of data, technicalities, morality and socio-political challenges. These challenges have to be solved by working out stringent requirements of data validation, adherence to moral standards, capacity-building, as well as cross-sectoral collaboration. Only by addressing these issues and harnessing the full capabilities of the geospatial technologies, researchers and policy makers can advance the more more subtle, fair, and efficient method of poverty reduction and eventually mitigate the socioeconomic inequalities and guarantee the establishment of more robust and diverse communities in the world.

REFERENCES

Alkire, S., & Foster, J. (2011). Counting and multidimensional poverty measurement. *Journal of Public Economics*, *95*(7–8), 476–487. DOI: 10.1016/j.jpubeco.2010.11.006

Barrington-Leigh, C. J., & Millard-Ball, A. (2017). The world's user-generated road map is more than 80% complete. *PLoS One*, *12*(8), e0180698. DOI: 10.1371/journal.pone.0180698 PMID: 28797037

Charmes, K. (1999). *Informal sector, poverty and gender: A review of empirical evidence. Background Paper for the World Development Report 2001*. World Bank.

De Sherbinin, A. (2018). The role of geospatial information in achieving the Sustainable Development Goals. *Geo-Spatial Information Science*, *21*(1), 21–30. DOI: 10.1080/10095020.2018.1498683

Duflo, E. (2012). Women empowerment and economic development. *Journal of Economic Literature*, *50*(4), 1051–1079. DOI: 10.1257/jel.50.4.1051

Elbers, B., Lanjouw, J., & Lanjouw, P. (2003). Micro-level estimation of poverty and inequality. *Econometrica*, *71*(1), 355–364. DOI: 10.1111/1468-0262.00399

Engstrom, R., Hersh, J. S., & Newhouse, D. (2017). Poverty from space: Using high-resolution satellite imagery for estimating economic well-being. *World Bank Policy Research Working Paper No. 8284*. DOI: 10.1596/1813-9450-8284

Feenstra, T. E., Inklaar, R., & Timmer, M. P. (2015). The next generation of the Penn World Table. *The American Economic Review*, *105*(10), 3150–3182. DOI: 10.1257/aer.20130954

Fields, A. (2004). Dualism in the labor market: A perspective on the Lewis model after half a century. *Manchester School*, *72*(6), 724–735. DOI: 10.1111/j.1467-9957.2004.00432.x

Floro, A. M. (1995). *Economic restructuring, gender and the allocation of time*. World Bank. DOI: 10.1016/0305-750X(95)00092-Q

Gibson, J., Olivia, S., & Rozelle, G. (2010). How deep is the depth of poverty? Evidence from Indonesia. *Asian Economic Journal, 24*(2), 99–123. DOI: 10.1111/j.1467-8381.2010.02031.x

International Labour Organization. (1998). *Women in the informal sector and their access to microfinance*. International Labour Organization.

International Labour Organization. (2012). *Rethinking economic growth: Towards productive and inclusive Arab societies*. International Labour Organization.

International Labour Organization. (2018). *Decent work and the Sustainable Development Goals: A guidebook on SDG labour market indicators*. International Labour Organization.

International Labour Organization. (2019). *Global wage report 2018/19: What lies behind gender pay gaps*. International Labour Organization.

International Labour Organization. (2020). *World employment and social outlook: Trends 2020*. International Labour Organization.

Jean, N., Burke, M., Xie, M., Davis, W. M., Lobell, D. B., & Ermon, S. (2016). Combining satellite imagery and machine learning to predict poverty. *Science, 353*(6301), 790–794. DOI: 10.1126/science.aaf7894 PMID: 27540167

Jerven, A. (2013). *Poor numbers: How we are misled by African development statistics and what to do about it*. Cornell University Press.

Kabeer, A. (2003). *Gender mainstreaming in poverty eradication and the Millennium Development Goals: A handbook for policy-makers and other stakeholders*. Commonwealth Secretariat. DOI: 10.14217/9781848598133-en

Lopez, R. A., & Valdes, A. (1997). Fighting rural poverty in Latin America: New evidence of the effects of education, demographics, and access to land. *Economic Development and Cultural Change, 45*(3), 487–507. DOI: 10.1086/452288

Organisation for Economic Co-operation and Development. (2019). *Society at a glance 2019: OECD social indicators*. OECD Publishing.

Ravallion, M. (2016). Poverty lines across the world. *The World Bank Research Observer, 31*(2), 259–287. DOI: 10.1093/wbro/lkw003

Smith, D. J., & Ward, M. W. (2014). Mapping poverty and inequality: New perspectives using GIS and remote sensing. *Computers, Environment and Urban Systems, 42*, 35–45. DOI: 10.1016/j.compenvurbsys.2013.10.001

Squire, L. M. (1996). Poverty in developing countries: Trends and policy responses. *The World Bank Research Observer, 11*(2), 153–171. DOI: 10.1093/wbro/11.2.153

Tatem, A., Campbell, C., & Guerra-Arias, D. (2014). Mapping for maternal and newborn health: The value of spatial data in health planning. *International Journal of Health Geographics, 13*(1), 1–9. DOI: 10.1186/1476-072X-13-13

Tokman, M. (1989). Integrating the informal sector into the modernization process. *World Development, 17*(7), 1065–1075. DOI: 10.1016/0305-750X(89)90113-1

Tokman, M. (1991). *Flexible work in Latin America.* International Labour Organization.

United Nations Development Programme. (2019). *Human development report 2019: Beyond income, beyond averages, beyond today—inequalities in human development in the 21st century*. UNDP.

United Nations Development Programme. (2020). *Human development report 2020: The next frontier—Human development and the Anthropocene*. UNDP.

United Nations Economic and Social Commission for Western Asia. (2014). *Arab middle class: Measurement and role in driving change*. UN ESCWA.

World Bank. (2013). *World development report 2013: Jobs.* World Bank.

World Bank. (2021). *Informality: Causes, consequences and policy responses.* World Bank.

Chapter 4
Policy, Ethics and Future Directions in Alleviating Poverty from Space

Soumen Ghosh
http://orcid.org/0009-0008-3778-6090
Global Institute of Management and Technology, Katwa, India

Rupam Hazra
http://orcid.org/0009-0003-0332-2128
Global Institute of Management & Technology, Ranaghat, India

ABSTRACT

Traditional poverty data in LMICs rely on infrequent surveys, leaving gaps for timely decisions. Space technologies such as Earth observation and mobile data provide scalable, high-frequency proxies to produce granular poverty maps between censuses. Features like night-time lights, roads, buildings, and agriculture correlate with welfare, and when fused with surveys through "data sandwich" methods, improve precision and coverage. Machine learning, especially CNNs, boosts predictive accuracy and ensures consistent estimates across scales. Policy, Ethics and Future Directions in Alleviating Poverty from Space highlights alignment of EO with SDG mandates, robust data governance, and collaboration across sectors. Ethical use requires privacy-by-design, fairness audits, transparency, and community engagement. Evidence from Africa, Mexico,

DOI: 10.4018/979-8-3373-7118-4.ch004

Bangladesh, Thailand, and Togo shows feasibility, accuracy, and integration into official statistics. Future directions stress multi-temporal monitoring, multimodal fusion, fairness-aware design, and EO-MLOps pipelines, as discussed in this chapter.

1. INTRODUCTION

Poverty remains a persistent challenge in low- and middle-income countries, where traditional measurement approaches rely on infrequent censuses and household surveys. These static methods often fail to capture rapid changes in welfare caused by migration, infrastructure development, or crises. Recent advances in space technologies, particularly Earth observation imagery, combined with mobile and survey data, have opened new possibilities for generating high-frequency, fine-grained poverty estimates. Machine learning models trained on multisource imagery can detect socioeconomic conditions through features such as night-time lights, road networks, and land use patterns, producing consistent estimates across scales. By blending surveys with space-enabled data, governments and researchers can improve precision, accelerate crisis response, and enhance program targeting. This chapter explores the policy frameworks, ethical safeguards, and future directions necessary to institutionalize space-enabled poverty alleviation.

2. BACKGROUND AND RATIONALE

Traditional poverty statistics in many LMICs rely on infrequent censuses and household surveys, leaving a persistent data gap that hampers timely decisions and monitoring of welfare dynamics. Earth observation imagery and anonymized mobile data now offer scalable, high-frequency proxies that fill these gaps with granular maps produced between censuses. Observable features—such as night-time lights, road networks, and proximity to urban Centers—correlate with socioeconomic conditions, enabling geospatial poverty estimation at fine spatial resolutions. Blending traditional surveys with satellite and mobile data through multimodal or "data sandwich" approaches improves precision, spatial coverage, and temporal relevance over single-source methods. Machine learning models trained on multi-source imagery can generate consistent estimates across scales from grid cells to administrative boundaries, enhancing within-country comparability for policy use. Governments and national statistical systems are beginning to institutionalize these tools; for example, Bangladesh's BBS reported union-level 2022 estimates using satellite imagery and ML with about 84% accuracy. Examining moral responsibility to society and to the impoverished is part of the ethical analysis of poverty. This includes talking about social justice, human rights, and the moral requirements for ending poverty by allocating opportunities and resources fairly (Jumayev, 2024).

A. Poverty Measurement Gaps and Data Scarcity

Poverty measurement in many low- and middle-income countries is limited due to infrequent household surveys and outdated census data. This creates a significant gap in understanding the real-time dynamics of poverty, making it difficult for policymakers to design targeted interventions. Traditional data collection methods are costly, time-consuming, and often fail to capture sudden economic shocks, such as those caused by pandemics or natural disasters. Furthermore, rural and marginalized communities are frequently underrepresented in official statistics, leading to incomplete poverty profiles. As a result, decision-makers face challenges in allocating resources effectively and monitoring progress toward poverty reduction goals. Nets and webs can be used to illustrate the multifaceted nature and connections between poverty and illness. Participatory approaches frequently surprise, expose, and shed light on aspects of poverty and illness that are often overlooked, such as seasonality, the locations of the impoverished, the significance of the body, and the ways in which these and other factors interact.

a) Infrequent and Outdated Data Collection

Most household surveys and censuses are conducted every 5–10 years, leaving long gaps in updated poverty information. This makes it hard to capture rapid changes in household income, migration patterns, or unemployment trends. The lack of timely data weakens the ability of governments and NGOs to respond to crises effectively.

b) High Cost and Logistical Challenges

Traditional data collection methods require extensive manpower, financial resources, and logistical planning, especially in remote or conflict-prone regions. These challenges often limit the frequency and quality of surveys. As a result, many countries struggle to sustain regular poverty monitoring initiatives.

c) Exclusion of Vulnerable Groups

Marginalized populations, including rural poor, informal workers, and displaced communities, are often excluded or underrepresented in surveys. This creates a biased picture of poverty, where the most vulnerable remain invisible in official statistics. Consequently, policies and aid programs risk overlooking those who need them most.

B. Evolution of Space-Based Socioeconomic Analytics

The use of satellite technology for socioeconomic analytics has transformed the way poverty, development, and human well-being are measured. Initially, satellites were primarily used for environmental monitoring, such as tracking deforestation or agricultural yields. Over time, researchers realized that features visible from space—like night-time lights, infrastructure density, and land-use patterns—could serve as proxies for economic activity. Recent advances in machine learning and AI have further enabled the extraction of socioeconomic indicators from high-resolution imagery with remarkable accuracy. Today, space-based analytics complement traditional surveys, offering scalable, cost-effective, and near real-time insights for policymakers and development agencies.

With the promise of scientific innovation and economic rejuvenation set against the backdrop of stark socioeconomic inequities, the growth of the space economy poses a contradiction with major societal implications (Koellner, 2024).

a) Shift from Environmental to Socioeconomic Applications

Early satellite missions focused mainly on climate, vegetation, and land mapping. However, researchers later identified correlations between satellite-derived features and human development metrics. This shift expanded the role of satellites from purely environmental tools to instruments for monitoring poverty, urban growth, and inequality.

b) Integration with AI and Big Data

Advances in artificial intelligence have made it possible to process massive volumes of satellite imagery efficiently. Machine learning algorithms can detect roads, housing patterns, and night-time lights to estimate income levels and population density. This integration has accelerated the accuracy and scalability of space-based socioeconomic assessments.

c) Support for Policy and Development Programs

Governments, NGOs, and global organizations now rely on satellite-based analytics to identify underserved regions and monitor progress toward sustainable development goals (SDGs). By providing up-to-date and granular data, these insights support better-targeted interventions. This has proven especially valuable in countries with scarce or outdated census information.

C. Integration with SDG 1 and Development Policy

The integration of advanced data tools with Sustainable Development Goal 1 (No Poverty) is crucial for tracking and accelerating progress toward eradicating extreme poverty. Traditional poverty measurements are often too slow to inform timely decisions, but space-based and digital analytics provide real-time insights that support evidence-based policymaking. By aligning poverty maps and socioeconomic indicators with SDG frameworks, governments can identify high-priority regions and allocate resources more effectively. This integration also enables continuous monitoring of poverty reduction programs, ensuring accountability and transparency in development efforts. Ultimately, linking innovative data systems with SDG 1 helps bridge the gap between global ambitions and local action.

To accomplish SDG 1, it is essential to implement coordinated and ongoing initiatives that tackle not just financial poverty but also the restricted availability of education, healthcare, social protection, and control over economic resources (Modise, 2025).

a) Targeted Resource Allocation

Space-based and digital poverty data help policymakers pinpoint regions with the highest deprivation. Instead of generalized programs, governments can channel resources to specific communities, maximizing impact. This targeted approach improves efficiency in implementing poverty reduction strategies aligned with SDG 1.

b) Monitoring and Evaluation of Progress

SDG 1 requires not only planning but also continuous measurement of outcomes. High-frequency data from satellites and mobile platforms allow governments and international bodies to evaluate whether poverty is decreasing in real time. This feedback loop ensures that interventions are adjusted when strategies fail to deliver expected results.

c) Evidence-Based Policy and Global Accountability

Integrating modern analytics with SDG tracking strengthens the credibility of development reporting. It reduces reliance on outdated or incomplete surveys and enhances global accountability mechanisms. This ensures that international donors, NGOs, and governments base their policies on reliable, transparent evidence, advancing both local and global poverty eradication agendas.

3. SPACE TECHNOLOGIES FOR POVERTY ALLEVIATION

Space technologies such as Earth observation now provide high-frequency, scalable data streams that generate granular poverty insights between surveys, strengthening monitoring and decision-making in data-scarce settings. Multispectral optical imagery from platforms like Landsat and Sentinel and night-time lights from VIIRS capture built environment, land use, and economic activity proxies that correlate with welfare, enabling spatially detailed mapping. High-resolution images allow extraction of features such as building density, road networks, roof materials, and agricultural patterns that predict local poverty at small administrative levels. Methodologically, feature-based, image-based, and hybrid approaches form the core toolbox for deriving poverty proxies from daytime and nighttime remote sensing. Advances in machine learning, particularly convolutional neural networks, improve predictive accuracy by learning representations directly from imagery and have been shown to outperform expert-defined feature methods. Night-time light–based models can produce internally consistent within-country poverty estimates across scales, from fine grids to administrative units, supporting comparable analysis and planning. Fusing satellite data with mobile phone metadata, surveys, and points of interest strengthens models and enables neighbourhood-level poverty estimation validated by ground truth. These space-enabled products support program targeting, resource allocation, and rapid crisis response by revealing evolving poverty hotspots when refreshed frequently. Leveraging economies of scale and the capability to overcome conventional limitations, space technology has the potential to significantly improve and broaden the overall efforts for alleviating poverty – impacting sectors like literacy, education, healthcare, inclusion, and societal empowerment (ESCAP, 2003).

A. Earth Observation: Optical, SAR, and Nighttime Lights

Earth observation technologies provide powerful tools for monitoring socioeconomic and environmental conditions at scale. Optical imagery captures high-resolution images of the Earth's surface, enabling analysis of infrastructure, agriculture, and land-use changes. Synthetic Aperture Radar (SAR) penetrates clouds and collects data day and night, offering reliable insights in regions with frequent cloud cover or poor visibility. Nighttime lights data measure human activity and energy consumption by detecting illumination, which strongly correlates with economic development. Together, these complementary methods provide a multi-dimensional understanding of poverty, urbanization, and development trends, bridging gaps left by traditional surveys.

a) Optical Imagery for Infrastructure and Land Use

Optical sensors capture detailed visual data on roads, housing density, and agricultural fields. These features serve as proxies for economic well-being, helping researchers identify areas with inadequate infrastructure. However, optical imagery depends on daylight and clear weather, which may limit its coverage in certain regions.

b) SAR for Reliable, All-Weather Observation

Unlike optical systems, SAR can penetrate clouds and operate at night, ensuring consistent monitoring. This makes it especially valuable in tropical and monsoon-prone regions where cloud cover is frequent. SAR data provide insights into terrain, building structures, and disaster impacts, offering critical input for poverty and resilience studies.

c) Nighttime Lights as a Proxy for Economic Activity

The brightness of night lights correlates with electricity access, industrial activity, and urban growth. Regions with weak or absent nighttime lights often overlap with poverty hotspots. Although not a perfect measure, nighttime light data offer cost-effective, near real-time indicators of economic development across countries and regions.

B. Geospatial and Mobile Data Fusion Approaches

The fusion of geospatial and mobile data has emerged as a powerful approach to measure and analyze socioeconomic conditions more accurately. Geospatial data, such as satellite imagery, provides macro-level insights into infrastructure, land use, and environmental patterns. Mobile data, including anonymized call records, mobility patterns, and mobile money transactions, captures micro-level human behavior and economic activity. When combined, these two data sources offer a holistic picture of both the physical environment and the socioeconomic dynamics of populations. This fusion enables more precise poverty mapping, disaster response planning, and policy targeting, particularly in regions where traditional surveys are scarce or outdated.

Both linear and polygonal data can be used with data fusion. Algorithms for connecting linear structures are introduced in this paper. Coordinate comparison is the foundation of the approach that is being explained (Stankutė & Asche, 2009).

a) Enhanced Poverty Mapping

By integrating satellite features (e.g., road density, vegetation health) with mobile phone usage patterns, researchers can estimate household wealth and access to services. This results in more accurate and granular poverty maps. Such fusion is especially valuable in rural or conflict-prone areas where survey data are missing.

b) Real-Time Human Mobility Insights

Mobile data reveal how populations move, migrate, or commute, while geospatial data show the conditions of the areas they move through. Together, they help track urbanization, labor migration, and the impact of crises like pandemics or natural disasters. This allows governments to plan transportation, housing, and social services more effectively.

c) Support for Targeted Policy and Aid Distribution

The fusion of mobile and geospatial data allows policymakers to identify communities most in need and deliver aid more efficiently. For example, mobile money data combined with geospatial poverty indicators can highlight households vulnerable to food insecurity. This ensures development funds and humanitarian aid reach the right beneficiaries at the right time.

C. Communications Satellites for Connectivity and Services

Communications satellites play a vital role in bridging the digital divide by extending internet and telecommunication services to underserved and remote regions. In many low- and middle-income countries, terrestrial infrastructure such as fiber-optic cables and cell towers is limited or expensive to deploy in rural areas. Satellites provide an alternative, offering wide-area coverage that can reach isolated communities. This connectivity supports access to education, healthcare, financial inclusion, and government services, all of which are critical for socioeconomic development. Moreover, satellite-based services enable disaster recovery communications when ground infrastructure is damaged, ensuring resilience in crises.

A new age of commercial space market prospects is being ushered in by the development of extremely powerful tiny satellites (Burleigh et al., 2019).

a) Expanding Rural and Remote Connectivity

Communications satellites deliver broadband internet and mobile networks to areas where laying cables is not feasible. This reduces digital exclusion and ensures that rural populations can participate in the digital economy. Access to reliable connectivity empowers local businesses, students, and farmers with new opportunities.

b) Enabling Essential Services

Satellite connectivity supports telemedicine, e-learning, mobile banking, and e-governance platforms. For instance, doctors can remotely diagnose patients, and students in rural schools can access online educational resources. These services directly contribute to poverty alleviation and align with development policy goals.

c) Disaster Response and Resilience

When earthquakes, floods, or storms disrupt terrestrial networks, satellites provide uninterrupted communication channels. Humanitarian agencies rely on satellite phones and internet links to coordinate relief operations. This resilience ensures that affected populations continue to receive aid and critical information during emergencies.

4. POLICY FRAMEWORKS AND GOVERNANCE

Effective frameworks for space-enabled poverty alleviation begin by aligning national plans and SDG monitoring mandates with Earth observation programs, clarifying institutional roles for data production, stewardship, and use. Governance should uphold peaceful use, non-appropriation, and equitable access to space-derived benefits, extending these principles to data-sharing and applications that prioritize vulnerable populations. Governments can codify data governance—access tiers, licensing, interoperability, and open-by-default geospatial standards—to maximize reuse of EO products while safeguarding quality and traceability. Regional cooperation and shared services reduce costs and strengthen resilience, with disaster risk reduction frameworks offering templates for cross-border EO data pipelines, common products, and capacity networks. Multi-stakeholder collaboration that links government, academia, and industry enables sustainable procurement, pilots-to-platforms scaling, and technology transfer for social protection and rural development use cases. Operational governance should require transparent model documentation, independent validation against ground truth, and periodic audits for bias and drift in EO- and ML-based poverty estimates. Due to the development of public health (PH) and the effects of globalization, this field has faced new moral dilemmas (Abbasi et al., 2018).

A. National Policy Alignment and Fiscal Linkages

National policy alignment with poverty reduction strategies is essential for ensuring that innovative data tools, like satellite analytics and mobile data, translate into actionable outcomes. Governments must integrate these insights into fiscal planning and development priorities to maximize their impact. Proper alignment ensures that data-driven poverty mapping informs budget allocations, infrastructure investments, and social protection programs. By linking fiscal policies with evidence-based indicators, resources can be directed toward regions and communities with the greatest needs. This integration strengthens accountability and ensures that development goals, including SDG 1, are effectively pursued at both national and local levels. The importance assigned to national policy goals, including those related to distribution, can vary from one country to another, while notable – and in certain nations, increasing – regional income inequalities may intensify some of the challenges and compromises brought about by decentralization (Joumard & Kongsrud, 2003).

a) Evidence-Based Budget Allocation

When poverty data is incorporated into fiscal planning, governments can channel funds more equitably. For example, regions identified as poverty hotspots through geospatial and survey data can receive targeted investments in health, education, and social safety nets. This reduces inefficiencies and ensures public spending achieves measurable outcomes.

b) Integration into National Development Frameworks

Aligning innovative poverty measurement with national strategies ensures consistency between global goals and domestic policies. This prevents fragmentation, making sure that data-driven insights directly shape programs such as rural electrification, agricultural subsidies, or urban housing schemes. Such integration creates a unified approach to poverty eradication.

c) Strengthening Transparency and Accountability

Linking fiscal policies with real-time data fosters accountability in how public funds are used. Citizens, NGOs, and oversight bodies can track whether budget allocations are reaching intended beneficiaries. This transparency builds trust in governance while promoting more inclusive and sustainable development.

B. International Coordination and Standards (UN, MDBs)

International coordination plays a central role in ensuring that poverty measurement and socioeconomic monitoring are consistent, reliable, and globally comparable. Organizations such as the United Nations (UN) and Multilateral Development Banks (MDBs), including the World Bank, IMF, and regional development banks, establish frameworks and standards that guide how data is collected, shared, and applied. These bodies promote harmonized methodologies, enabling cross-country comparisons and supporting global goals like the Sustainable Development Goals (SDGs). They also provide technical and financial support to low- and middle-income countries that lack the capacity to implement advanced data systems. By setting standards and fostering collaboration, international institutions ensure that innovative tools like satellite analytics and mobile data contribute effectively to global development policy. Multilateral development banks (MDBs) implement environmental and social safeguard frameworks (ESFs) along with procurement policies to maintain high operational standards and minimize unintentional adverse effects (Getzel & Humphrey, 2024).

a) Harmonization of Methodologies

The UN and MDBs develop standardized guidelines for poverty measurement and socioeconomic data collection. This ensures that indicators such as income levels, access to services, and multidimensional poverty can be compared across countries. Harmonization prevents fragmented approaches and strengthens the credibility of global development reporting.

b) Capacity Building and Technical Assistance

Many low-income countries lack the technical expertise or infrastructure to adopt advanced geospatial and mobile data tools. MDBs and UN agencies provide training, funding, and technical support to help these countries modernize their statistical systems. This builds long-term capacity for evidence-based policymaking.

c) Global Accountability and Resource Mobilization

International organizations use standardized data to monitor progress toward SDGs, enabling transparent global accountability. MDBs also leverage this data to guide lending decisions, ensuring resources are directed toward countries and sectors with the greatest needs. This coordination strengthens trust among donors, governments, and civil society.

C. Data Governance: Access, Licensing, and Public Goods

Data governance is critical to ensuring that innovative poverty measurement tools—such as Earth observation, mobile data, and big data analytics—are used responsibly and equitably. Clear frameworks around access, licensing, and ownership determine who can use data, under what conditions, and for what purposes. Without proper governance, valuable socioeconomic data may remain locked in private silos, limiting its utility for public policy and development. Treating certain datasets, such as poverty maps and basic geospatial indicators, as public goods allows governments, researchers, and civil society to use them for inclusive development. At the same time, governance frameworks must balance openness with concerns around privacy, security, and commercial interests. Public goods are characterized by being non-rivalrous and non-excludable (Purtova & van Maanen, 2024).

a) Access and Licensing Frameworks

Data access is often restricted by commercial licenses or institutional agreements. Establishing open-access models for essential datasets—such as poverty indicators and satellite imagery—ensures wider usability. Flexible licensing arrangements can balance private sector interests with the broader public need for reliable, affordable data.

b) Data as a Public Good

Treating foundational data as a public good ensures that critical resources, like poverty maps, climate risk layers, and basic demographic indicators, are available to all. This approach prevents exclusion, promotes equity, and supports collaborative problem-solving among governments, NGOs, and researchers. Public good framing also helps justify public investments in data infrastructure.

c) Balancing Privacy, Security, and Utility

While open data fosters innovation, sensitive information—such as anonymized mobile phone records—requires strong governance to protect privacy. Regulations must establish safeguards to prevent misuse while still enabling valuable research and policymaking. Achieving this balance is key to building public trust in data-driven development.

5. ETHICS AND RESPONSIBLE USE

Ethical and responsible use of space-enabled poverty analytics begins with rigorous privacy-by-design, including data minimization, aggregation, and reidentification risk assessments when combining satellite imagery with mobile or survey data, centering privacy as a non-negotiable safeguard. Informed consent, purpose limitation, and clear public communication should govern the collection and linkage of auxiliary datasets and ground-truth surveys so communities understand potential impacts on eligibility and services. Ensuring fairness requires diagnosing and mitigating systematic errors such as urban–rural performance gaps and mis-ranking that can skew resource allocation and invisible vulnerable groups. Transparent model documentation, data lineage, and uncertainty/error maps enable external scrutiny and help policymakers interpret confidence before using outputs for targeting. Robust validation protocols should include independent audits, cross-context generalization tests, and continuous drift monitoring as landscapes, lighting, and economic signals change over time.

Explainability and interpretable features—alongside post hoc tools—are necessary so decisions are not driven by black-box indicators that may embed spurious correlations or hidden biases. Community engagement and accessible grievance redress mechanisms respect dignity, surface local definitions of deprivation, and allow rapid correction of harmful model impacts. The adoption of our approach to promote ethical decision-making will enhance suitability and usability, which are just two facets of policy quality (Ruighaver et al., 2010).

A. Privacy, Consent, and Re-identification Risks

The rapid expansion of geospatial, mobile, and big data analytics in poverty measurement introduces significant concerns around privacy and consent. While datasets are often anonymized, individuals can sometimes be re-identified when multiple data sources are combined, especially in small communities or unique mobility patterns. This creates risks of surveillance, misuse, or unintended harm to vulnerable populations. In many low- and middle-income countries, informed consent frameworks for large-scale data collection remain weak, leaving citizens unaware of how their information is being used. Without robust privacy safeguards, trust in data-driven development initiatives may erode, limiting their long-term effectiveness.

Numerous healthcare organizations adhere to data protection guidelines that outline which patient identifiers need to be removed in order to share "de-identified" records. However, these guidelines are frequently implemented without an understanding of the potential risk of "re-identification"(Benitez & Malin, 2010).

a) Challenges in Ensuring Informed Consent

Many poverty data sources—like mobile call detail records (CDRs) or satellite-derived indicators—are collected passively without direct individual consent. Populations often lack clarity on how their data is being used or shared. Strengthening informed consent practices is vital to uphold ethical standards and public trust.

b) Risks of Re-identification

Even anonymized datasets can be vulnerable to re-identification when cross-linked with other information, such as voter registries, social media, or location patterns. This can expose sensitive details about individuals or communities, leading to risks like discrimination or targeted exploitation. Researchers must apply advanced privacy-preserving methods to reduce this danger.

c) Balancing Data Utility and Privacy Protection

Excessive restrictions on data use can limit innovation and the usefulness of socioeconomic analytics, while weak protections can compromise individual rights. Governance frameworks need to strike a balance by adopting tools like differential privacy, federated learning, and strict access controls. This balance ensures data can inform policy without undermining personal freedoms.

B. Bias, Fairness, and Representativeness in Models

The use of AI, satellite imagery, and mobile data in poverty measurement raises important concerns about bias, fairness, and representativeness. Data-driven models are only as good as the data they are trained on; if marginalized communities are underrepresented, the resulting insights risk reinforcing inequality. For example, rural areas with poor connectivity may generate fewer mobile data, making them appear less significant in poverty maps. Similarly, satellite imagery may fail to capture informal settlements hidden under tree cover or mixed-use land patterns. Ensuring fairness means designing models that do not disproportionately Favor urban, wealthy, or digitally connected populations at the expense of the poor and invisible. A key aspect of data quality is ensuring it represents all potential subgroups affected by that decision (Rodolfa et al., 2020).

a) Data Coverage Bias

Certain groups—such as rural poor, women without mobile phones, or displaced populations—are often missing from large datasets. Models trained on incomplete or skewed data may systematically underrepresent these groups. This results in policies that overlook the very populations most in need.

b) Algorithmic Fairness and Equity

Machine learning algorithms can unintentionally amplify social and economic disparities if not designed with fairness in mind. For instance, over-reliance on night-time lights as a poverty proxy may undervalue regions where electricity is scarce but poverty is high. Incorporating fairness checks helps ensure models benefit all groups equitably.

c) Improving Representativeness Through Hybrid Approaches

Combining traditional household surveys with geospatial and mobile data improves representativeness. Surveys provide ground truth for marginalized groups, while digital data scales insights across regions. This hybrid approach reduces bias, enhances fairness, and ensures development interventions reach the right populations.

C. Community Engagement and Lived-Space Considerations

While advanced technologies like satellites, mobile data, and AI provide valuable socioeconomic insights, they must be grounded in community engagement and lived-space perspectives. Poverty is not only a statistical condition but also a lived experience shaped by cultural, social, and local contexts. Top-down approaches that rely solely on remote sensing or algorithmic models risk misrepresenting or overlooking community realities. Engaging local populations ensures that poverty metrics reflect actual needs and priorities, while also building trust in data-driven policies. Incorporating lived-space knowledge alongside technological tools creates a more inclusive, human-centered framework for poverty measurement and development planning. Over time, the methods of social production of space influence the lived experience through cognitive, structural, and financial dimensions of spatial existence (Daradkeh, 2023).

a) Participatory Data Practices

Involving communities in data collection and validation ensures that indicators reflect local realities. For example, residents can provide insights into informal work, housing quality, or access to services that satellites cannot capture. Participatory practices empower communities and improve the accuracy of poverty assessments.

b) Respecting Cultural and Spatial Contexts

Different communities define and experience poverty in unique ways. For instance, access to grazing land may be critical for pastoralist groups, while urban slums face challenges of overcrowding and sanitation. Recognizing these differences ensures that data interpretations and policies are context-sensitive rather than one-size-fits-all.

c) Building Trust and Ownership

Transparent engagement fosters trust between governments, researchers, and local populations. When communities feel ownership of the data process, they are more likely to support policy interventions. This bottom-up approach makes poverty reduction strategies more sustainable and responsive to lived realities.

6. EVIDENCE AND CASE STUDIES

Foundational evidence shows that combining high-resolution daytime satellite imagery with machine learning can accurately predict local consumption and wealth across multiple African countries, demonstrating feasibility and scalability for poverty estimation. Complementary studies integrating mobile phone metadata with satellite data have produced accurate poverty maps, underscoring the value of multimodal data for welfare measurement in low- and middle-income settings. In Mexico, small area estimation that blends satellite imagery with household surveys improved the precision and accuracy of municipal poverty rates relative to survey-only approaches when census data were outdated. Urban-focused applications show neighbourhood-level poverty estimation from satellite imagery can inform service delivery and planning, as demonstrated in an IFC study at sub-city scales. In Thailand, an ADB-backed initiative mapped the spatial distribution of poverty using satellite Earth observation to guide regional development priorities. Institutional adoption is growing; Bangladesh's Bureau of Statistics produced 2022 union-level poverty estimates using satellite imagery and ML with a reported 84% accuracy, illustrating integration into official statistics. During COVID-19, Togo's Novissi program used satellite and mobile phone data with machine learning to target emergency cash transfers, reducing exclusion errors by 4–21% compared to geographic blanketing while maintaining fairness across groups. Policy development is frequently mentioned as a key role of Hospital Ethics Committees (HECs), alongside consultation and education (Frolic & Drolet, 2013).

A. Poverty Mapping with Satellite and Mobile Data

The integration of satellite imagery and mobile phone data has transformed how poverty is measured and mapped, especially in regions where surveys are infrequent or outdated. Satellite data, such as night-time lights, land use, and infrastructure patterns, provide proxies for economic activity and development. Mobile data, including call detail records, mobility trends, and mobile money transactions, capture human behavior and access to services at a granular level. When combined with machine learning models, these datasets produce highly detailed poverty maps that reveal variations across neighbourhoods, villages, or districts. Such maps allow governments, NGOs, and development agencies to target interventions more effectively and monitor progress toward poverty reduction goals in near real time. Poverty is a significant factor influencing negative health outcomes worldwide, a primary contributor to social instability, and one of the foremost reasons for the loss of human potential (Steele et al., 2017).

a) Satellite Data as Poverty Proxies

High-resolution imagery helps identify features like road networks, housing quality, and agricultural land. Night-time lights, for instance, often correlate with electricity access and income levels. These proxies allow researchers to estimate poverty even in remote or inaccessible regions.

b) Mobile Data for Behavioral Insights

Anonymized mobile phone records provide information on communication patterns, mobility, and financial inclusion. For example, frequent mobile money transactions may indicate better economic activity, while limited mobility may signal isolation and deprivation. These insights complement physical indicators from satellites.

c) Fusion and Machine Learning for Granular Mapping

Combining satellite and mobile datasets using AI models produces fine-grained poverty maps with higher accuracy than single-source approaches. These maps reveal intra-regional disparities—such as poverty pockets within otherwise wealthy districts—helping policymakers design targeted social programs.

B. Nighttime Lights for Targeting and Monitoring

Nighttime light (NTL) imagery from satellites has become one of the most widely used proxies for socioeconomic development. The intensity and distribution of artificial lighting at night correlate strongly with electricity access, urbanization, and economic activity. Policymakers and researchers use NTL data to identify regions lacking infrastructure, target aid programs, and monitor development progress over time. Because these datasets are updated frequently and cover nearly the entire globe, they provide a cost-effective alternative to traditional surveys, especially in countries with scarce or outdated census data.

However, while powerful, NTL data must be combined with other indicators to avoid misinterpretations in rural, low-light, or culturally distinct contexts. While both have been essential for ALAN research, the sensing of nighttime lights was not the main design goal for either the DMSP-OLS or VIIRS-DNB instruments; consequently, they possess important limitations (Barentine et al., 2021).

a) Targeting Underserved Regions

Areas with weak or no night lighting often overlap with communities lacking electricity, roads, or healthcare services. Governments and aid organizations can use this information to prioritize infrastructure investments. This targeting ensures that scarce resources reach the most deprived populations.

b) Monitoring Development Progress

Changes in nighttime light intensity over time can signal economic growth, electrification, or disaster recovery. For example, post-disaster recovery efforts can be tracked by observing how quickly light levels return to normal. This allows continuous and objective monitoring of development initiatives.

c) Limitations and Complementary Data Needs

While useful, NTL imagery may underrepresent rural poverty where communities rely less on electricity or use off-grid renewable systems. Similarly, brightly lit industrial zones may give the impression of wealth despite local poverty. Combining NTL with survey data, mobile records, or SAR imagery provides a more accurate and balanced picture.

C. Program Design and Evaluation Enabled by EO

Earth Observation (EO) technologies provide policymakers and development agencies with powerful tools for designing, implementing, and evaluating poverty reduction programs. By capturing high-resolution data on land use, infrastructure, environmental conditions, and human activity, EO helps identify priority areas for intervention. For example, satellite imagery can highlight regions with poor road access, vulnerable agricultural zones, or informal settlements that require targeted support. During implementation, EO enables continuous monitoring of program reach and impact in near real time, reducing dependence on infrequent surveys. In the evaluation phase, EO data allows for objective measurement of progress, ensuring transparency, accountability, and evidence-based adjustments to policies and projects. Satellite Earth observation provides ample data, and thus has provided opportunities for the development of novel products and services with the potential to support implementation, monitoring and reporting for AU-Agenda 2063 development imperatives (Kganyago & Mhangara, 2019).

a) Informed Program Targeting

EO helps governments and NGOs identify poverty hotspots and resource gaps. For instance, mapping areas without access to electricity or safe water can guide the placement of infrastructure projects. This ensures that program design is based on objective, spatially detailed evidence.

b) Monitoring Implementation in Real Time

With frequent satellite updates, development agencies can track whether infrastructure projects—such as schools, clinics, or irrigation systems—are reaching intended areas. This reduces delays, detects leakages or inefficiencies, and improves accountability throughout program execution.

c) Evaluation and Impact Assessment

EO provides independent, verifiable data to assess whether interventions are reducing poverty and improving livelihoods. For example, vegetation indices can measure agricultural productivity gains from subsidy programs, while nighttime lights can track electrification impacts. These evaluations strengthen credibility and guide future policy refinements.

7. RISK MANAGEMENT AND SAFEGUARDS

In the field of social work, the term 'risk' is generally interpreted as the potential for negative outcomes, particularly those of a severe nature (Macdonald & Macdonald, 2010). Effective risk management for space-enabled poverty analytics starts with privacy-by-design—data minimization, aggregation, impact assessments, and reidentification testing—before linking satellite, mobile, and survey data to prevent harm. Consent practices, purpose limitation, and lawful processing should govern auxiliary data collection and linkage, supported by access tiers, retention limits, and clear licensing to control sensitive use. Model risk management must include systematic bias audits across urban–rural and other relevant subgroups, out-of-context generalization checks, and routine drift monitoring as environments and economies evolve. Uncertainty quantification, calibration, and public error/coverage maps are essential so decision-makers can interpret confidence intervals before high-stakes targeting or evaluation. Transparent documentation of data lineage, features, and model architectures—plus independent replication and red-teaming—builds trust and enables rigorous external scrutiny. Human-in-the-loop review and ethical oversight boards should gate eligibility or benefits decisions, ensuring model indicators complement (not replace) needs assessments and social registries. Security safeguards must protect geospatial data lakes and model artifacts against unauthorized access, with incident response playbooks for misuse, erroneous outputs, and unintended impacts on vulnerable communities.

A. Ethical Review and Data Protection Protocols

As Earth Observation (EO), mobile, and big data technologies become central to poverty mapping and development planning, ethical review and data protection protocols are critical to safeguard individual rights and community trust. These protocols ensure that data collection, processing, and sharing are conducted responsibly, without exposing vulnerable populations to harm. Ethical reviews—often guided by institutional review boards (IRBs) or data ethics committees—evaluate whether projects respect principles of privacy, consent, and fairness. Meanwhile, data protection protocols establish clear rules for storage, access, and usage, reducing risks of misuse or unauthorized re-identification. Embedding ethics into data-driven development strengthens legitimacy and ensures that innovation aligns with human rights standards. The correct interpretation of, and adherence to, data protection and freedom of information laws concerning various administrative and educational functions carried out in a UK higher education institution continues to be a source of conflict, confusion, and poorly designed institutional policies more than ten years after the legislation was enacted (Charlesworth, 2012).

a) Institutional Ethical Oversight

Formal ethical review processes, such as IRBs, provide independent assessments of data projects before implementation. They examine potential risks to individuals and communities, ensuring that benefits outweigh harms. This oversight fosters accountability and consistency across research and policy applications.

b) Robust Data Protection Protocols

Secure storage, encryption, and controlled access mechanisms are essential to prevent unauthorized use of sensitive datasets. Protocols must define who can access data, under what conditions, and for how long. Adopting international standards (e.g., GDPR principles) helps harmonize practices across borders.

c) Safeguarding Vulnerable Populations

Ethical data governance prioritizes the protection of marginalized groups, who are often most at risk of exploitation. By applying techniques like anonymization, differential privacy, or federated learning, researchers can balance data utility with strong safeguards. This ensures that development initiatives empower rather than endanger communities.

B. Model Risk, Drift, and Monitoring Frameworks

As AI and machine learning models are increasingly used for poverty mapping and socioeconomic analysis, they bring new challenges related to model risk, drift, and monitoring. Model risk refers to the possibility that algorithms produce inaccurate, biased, or misleading outputs that negatively impact decision-making. Over time, models are also subject to drift—where the data environment changes (e.g., migration, new infrastructure, or economic shocks) and the model's predictions lose accuracy. To address these issues, continuous monitoring frameworks are essential to evaluate model performance, detect anomalies, and update algorithms as conditions evolve. Without strong governance and oversight, model errors could lead to misallocation of resources or reinforce inequalities.

a) Understanding Model Risk

Model risk arises when predictive algorithms are trained on incomplete, outdated, or biased data. For example, over-reliance on night-time lights may overlook off-grid communities, leading to skewed poverty maps. Recognizing and quantifying model risk is the first step toward responsible deployment.

b) Addressing Data and Concept Drift

As socioeconomic conditions shift, models may no longer reflect reality. For instance, mobile phone usage patterns can change after a pandemic or new telecom policy. Monitoring for data drift (changes in input data) and concept drift (changes in relationships between inputs and outcomes) ensures models remain reliable over time.

c) Implementing Monitoring and Governance Frameworks

Regular audits, performance tracking, and human oversight are crucial for maintaining model integrity. Frameworks should include feedback loops from field surveys and communities to validate predictions. This hybrid monitoring approach strengthens accountability and minimizes risks of harmful misclassifications.

C. Conflict Sensitivity and Do-No-Harm Principles

The use of satellite imagery, mobile data, and AI for poverty measurement must be guided by conflict sensitivity and do-no-harm principles. In fragile and conflict-affected settings, poorly designed data initiatives can unintentionally worsen tensions, exclude vulnerable groups, or expose sensitive information that could be misused by state or non-state actors. For example, detailed poverty maps might reveal the locations of displaced populations or marginalized minorities, making them targets in ongoing conflicts. Applying conflict sensitivity means recognizing the political and social dynamics of data use, ensuring that interventions reduce, rather than exacerbate, risks. Grounding poverty measurement in do-no-harm principles ensures that data-driven development empowers communities and contributes to peacebuilding.

a) Recognizing Political and Social Dynamics

Data is never neutral—it can influence power balances, resource allocation, and perceptions of inequality. Conflict-sensitive approaches analyze how poverty data might interact with existing tensions, ensuring that results are not weaponized or misinterpreted.

b) Protecting Vulnerable Populations

Data initiatives must prioritize the safety of conflict-affected groups. This requires strict anonymization, careful sharing protocols, and engagement with humanitarian actors to avoid exposing sensitive population details. Protecting at-risk groups prevents harm while still enabling valuable analysis.

c) Embedding Do-No-Harm in Program Design

Poverty measurement should actively consider unintended consequences during design, implementation, and dissemination. Applying do-no-harm frameworks involves testing scenarios, consulting local stakeholders, and adapting outputs to minimize risks. This ensures data contributes to resilience and stability rather than fueling mistrust or division.

8. FUTURE DIRECTIONS

The next phase of space-enabled poverty analytics will emphasize multi-temporal, fine-scale estimation by integrating new EO missions (e.g., SDGSAT-1) with functional zoning and time-series signals to track change continuously. Advances in deep learning—especially self-supervised CNNs and sequence models—will extract robust features from minimal labels and capture seasonal and structural dynamics in welfare proxies. A shift toward truly multimodal fusion that blends satellite imagery with mobile phone data, points of interest, and other geospatial covariates will improve generalization and stability across diverse contexts. Deployment practice will center on calibrated uncertainty, coverage diagnostics, and public error maps so outputs are decision-ready, not just predictive. Fairness-aware design with subgroup performance reporting and bias audits will become standard to prevent systematic misclassification of rural, informal, or marginalized communities. Production EO-MLOps—automated ingestion, change detection, and periodic retraining—will enable quarterly refreshes and rapid crisis tasking for social protection and disaster response. Institutionalization will deepen as national statistical systems embed EO in official statistics and SDG workflows, evidenced by Bangladesh's union-level adoption and SDG-focused satellite missions.

A. Next-Gen Sensors, ML, and Multimodal Fusion

The future of poverty measurement and socioeconomic monitoring is being reshaped by next-generation sensors, machine learning (ML), and multimodal data fusion. Advanced satellite sensors now provide ultra-high-resolution imagery, hyperspectral data, and continuous global coverage, capturing socioeconomic conditions with unprecedented detail. ML techniques enable efficient processing of these massive datasets, extracting patterns from roads, housing quality, crop yields, and mobility behaviours. Multimodal fusion—integrating EO data, mobile phone usage, social media, and administrative records—offers a holistic understanding of poverty dynamics that no single source could provide. Together, these technologies promise real-time, hyperlocal poverty analytics that can better guide development policies and humanitarian interventions.

a) Advancements in Next-Gen Sensors

New optical, SAR, and hyperspectral sensors provide richer data on land, infrastructure, and environmental conditions. For example, hyperspectral imagery can detect crop stress or water scarcity, offering indirect insights into household vulnerability. These sensors expand the scope of poverty-relevant indicators.

b) Machine Learning for Scalable Insights

ML algorithms process terabytes of imagery and Behavioral data to uncover complex correlations between observable features and poverty outcomes. Techniques like deep learning enable automated detection of informal settlements, transport networks, or electricity access, scaling analysis across entire countries.

c) Multimodal Fusion for Contextual Accuracy

Combining diverse data streams—satellites, mobile records, surveys, and even social media—reduces bias and improves representativeness. For instance, survey "ground truth" can validate satellite and mobile models, while mobile money data complements EO-based infrastructure insights. This multimodal approach creates richer, fairer, and more actionable poverty maps.

B. Responsible Commercialization and Policy Gaps

The growing role of private companies in Earth Observation (EO), mobile data, and AI-driven poverty mapping raises important questions about responsible commercialization and policy gaps. While the private sector provides cutting-edge technologies, scalable platforms, and innovative solutions, its profit motives may conflict with the public good. Many critical datasets remain locked behind paywalls or restrictive licenses, limiting their accessibility for governments, NGOs, and researchers in low-resource settings. Policy frameworks have not kept pace with these developments, leaving gaps around pricing models, equitable access, and ethical use of commercially provided data. Bridging these gaps requires a balance between incentivizing private innovation and ensuring that essential poverty-related data is treated as a global public good.

a) Equitable Data Access and Affordability

Commercial providers often price high-resolution satellite imagery and analytics beyond the reach of developing countries. Without policies ensuring affordability, data-driven poverty solutions risk deepening inequalities. Equitable licensing models and tiered pricing can help expand access.

b) Ethical Standards for Private Sector Involvement

Companies collecting and analyzing poverty-related data must adhere to ethical standards around privacy, consent, and fairness. Clear guidelines ensure that commercial innovation does not compromise vulnerable populations. Responsible commercialization requires stronger alignment with international human rights principles.

c) Policy and Regulatory Gaps

Current governance structures lag behind rapid technological advances. There is limited regulation on data monopolies, fair competition, or accountability mechanisms for misuse of poverty analytics. Closing these gaps will require international coordination, public–private partnerships, and updated legal frameworks.

C. Financing, Aid Effectiveness, and Sustainable Scaling

The successful adoption of advanced poverty measurement technologies depends not only on innovation but also on financing strategies, aid effectiveness, and sustainable scaling. Investments in satellite systems, data analytics platforms, and AI tools require significant upfront costs, which many low- and middle-income countries cannot bear alone. Donor funding, blended finance, and public–private partnerships are critical for providing the necessary resources. Moreover, linking these investments to measurable development outcomes ensures aid effectiveness, allowing donors and governments to track the impact of interventions. Sustainable scaling requires building local technical capacity, integrating tools into national systems, and creating governance frameworks that maintain long-term operational viability.

a) Strategic Financing for Data-Driven Development

High-resolution imagery, mobile analytics platforms, and AI tools require significant capital investment. Blended finance models—combining public, private, and donor funds—can lower barriers to entry and ensure that countries benefit from cutting-edge technology without overburdening national budgets.

b) Maximizing Aid Effectiveness

Integrating real-time poverty data with program planning allows donors and governments to direct funds where they are most needed. Evidence-based allocation improves transparency, accountability, and measurable impact, making aid more effective and efficient.

c) Sustainable Scaling and Local Capacity Building

Technologies must be embedded within local institutions to ensure long-term sustainability. This includes training local analysts, transferring technical knowledge, and integrating data systems into national planning frameworks. Sustainable scaling ensures that poverty measurement continues to inform policy even after initial donor or private-sector support ends.

9. CONCLUSION

Space technologies are reshaping how poverty is measured and addressed by providing scalable, timely, and granular insights that complement traditional surveys. Evidence from Africa, Mexico, Bangladesh, Thailand, and Togo illustrates both feasibility and growing institutional adoption of Earth observation and multimodal analytics. However, responsible use depends on robust governance frameworks, privacy safeguards, transparency, and community engagement to prevent bias or misuse. Looking forward, advances in multi-temporal monitoring, multimodal fusion, fairness-aware design, and EO-MLOps pipelines will strengthen the reliability and policy relevance of these tools. By embedding such practices into official statistics and SDG monitoring, space-enabled poverty analytics can drive more responsive, equitable, and sustainable development strategies, as highlighted throughout this chapter.

REFERENCES

Abbasi, M., Majdzadeh, R., Zali, A., Karimi, A., & Akrami, F. (2018). The evolution of public health ethics frameworks: Systematic review of moral values and norms in public health policy. *Medicine, Health Care, and Philosophy*, *21*(3), 387–402. DOI: 10.1007/s11019-017-9813-y PMID: 29124449

Barentine, J. C., Walczak, K., Gyuk, G., Tarr, C., & Longcore, T. (2021). A case for a new satellite mission for remote sensing of night lights. *Remote Sensing (Basel)*, *13*(12), 2294. DOI: 10.3390/rs13122294

Benitez, K., & Malin, B. (2010). Evaluating re-identification risks with respect to the HIPAA privacy rule. *Journal of the American Medical Informatics Association : JAMIA*, *17*(2), 169–177. DOI: 10.1136/jamia.2009.000026 PMID: 20190059

Burleigh, S. C., De Cola, T., Morosi, S., Jayousi, S., Cianca, E., & Fuchs, C. (2019). From Connectivity to advanced internet services: A comprehensive review of small satellites communications and networks. *Wireless Communications and Mobile Computing*, *2019*(1), 6243505. DOI: 10.1155/2019/6243505

Charlesworth, A. (2012). Data protection, freedom of information and ethical review committees: Policies, practicalities and dilemmas. *Information Communication and Society*, *15*(1), 85–103. DOI: 10.1080/1369118X.2011.637572

Daradkeh, L. (2023). Refugee camp streets: Unraveling lived space and community empowerment-a comparative case study analysis (Doctoral dissertation, Cardiff University).

ESCAP, U. (2003). Study on the use of space technology applications for poverty alleviation: trends, strategies and policy frameworks.

Frolic, A. N., & Drolet, K. (2013). Ethics policy review: A case study in quality improvement. *Journal of Medical Ethics*, *39*(2), 98–103. DOI: 10.1136/medethics-2011-100461 PMID: 23118469

Getzel, B., & Humphrey, C. (2024). *Streamlining MDB environmental, social and procurement safeguards while maintaining high standards.* ODI.

Joumard, I., & Kongsrud, P. M. (2003). Fiscal relations across government levels. *OECD Economic Studies, 2003*(1), 155–229.

Jumayev, O. M. (2024). PHILOSOPHICAL ISSUES OF POVERTY: SOCIAL, ETHICAL, AND POLITICAL ASPECTS. European science review, (5-6), 41-49.

Kganyago, M., & Mhangara, P. (2019). The role of African emerging space agencies in earth observation capacity building for facilitating the implementation and monitoring of the African development agenda: The case of African earth observation program. *ISPRS International Journal of Geo-Information, 8*(7), 292. DOI: 10.3390/ijgi8070292

Koellner, E. (2024). Navigating the New Frontier: Ethical, Societal, and Legal Challenges in the Space Economy. In *AIAA Aviation Forum and Ascend 2024* (p. 4825).

Macdonald, G., & Macdonald, K. (2010). Safeguarding: A case for intelligent risk management. *British Journal of Social Work, 40*(4), 1174–1191. DOI: 10.1093/bjsw/bcq041

Modise, J. M. (2025). ERADICATING POVERTY IN ALL ITS DIMENSIONS: A POLICY-ORIENTED APPROACH TO ACHIEVING SUSTAINABLE DEVELOPMENT GOAL.

Purtova, N., & van Maanen, G. (2024). Data as an economic good, data as a commons, and data governance. *Law, Innovation and Technology, 16*(1), 1–42. DOI: 10.1080/17579961.2023.2265270 PMID: 39634789

Rodolfa, K. T., Saleiro, P., & Ghani, R. (2020). Bias and fairness. In *Big data and social science* (pp. 281–312). Chapman and Hall/CRC. DOI: 10.1201/9780429324383-11

Ruighaver, A. B., Maynard, S. B., & Warren, M. (2010). Ethical decision making: Improving the quality of acceptable use policies. *Computers & Security, 29*(7), 731–736. DOI: 10.1016/j.cose.2010.05.004

Stankutė, S., & Asche, H. (2009, June). An integrative approach to geo-spatial data fusion. In *International Conference on Computational Science and Its Applications* (pp. 490-504). Berlin, Heidelberg: Springer Berlin Heidelberg. DOI: 10.1007/978-3-642-02454-2_35

Steele, J. E., Sundsøy, P. R., Pezzulo, C., Alegana, V. A., Bird, T. J., Blumenstock, J., Bjelland, J., Engø-Monsen, K., de Montjoye, Y.-A., Iqbal, A. M., Hadiuzzaman, K. N., Lu, X., Wetter, E., Tatem, A. J., & Bengtsson, L. (2017). Mapping poverty using mobile phone and satellite data. *Journal of the Royal Society, Interface*, *14*(127), 20160690. DOI: 10.1098/rsif.2016.0690 PMID: 28148765

Chapter 5
Spatial-Fuzzy Integration for Poverty Response:
Mapping Uncertainty, Vulnerability, and Policy Action

D. Rajalakshmi
http://orcid.org/0000-0003-4851-7685
SASTRA (Deemed to be) University, Thanjavur, India

Durga karthik
http://orcid.org/0000-0003-3199-8814
Srinivasa Ramanujan Centre, SASTRA (Deemed to be) University, Thanjavur, India

ABSTRACT

Working poverty, where individuals remain below poverty thresholds despite employment poses complex challenges that are deeply embedded in spatial and socio-economic contexts. Traditional statistical and econometric tools often struggle to capture the uncertainty, vagueness, and multidimensionality inherent in poverty indicators such as income stability, housing quality, healthcare access, and transport connectivity. This chapter introduces fuzzy soft sets as a novel mathematical framework for spatial poverty, the approach enables nuanced classification of poverty

DOI: 10.4018/979-8-3373-7118-4.ch005

zones, flexible handling of incomplete or imprecise data, and multi-criteria decision support for resource allocation. Case illustrations demonstrate how fuzzy soft sets can model overlapping poverty clusters, rank vulnerable districts for targeted interventions, and enhance spatial dashboards for policymakers. The chapter argues that fuzzy soft sets not only enrich the methodological toolkit for poverty studies but also foster interdisciplinary dialogue between mathematics, social science, and public policy.

INTRODUCTION

Despite considerable efforts, working poverty remains as perplexing as it is challenging. Pervasive working poverty is peculiar, as employing more people experiencing poverty is unlikely to address it. Instead, working poverty is more symptomatic of structural issues in economic systems: low wages, insecure employment, increasing prices of basic services, and unequal geographical distribution of economic, social, and civic opportunities. Millions of workers, both in developed and developing economies, find themselves in a situation in which full-time employment still keeps them in material deprivation, social exclusion, and economic vulnerability.

To understand and tackle working poverty, one needs to deal with many analytical obstacles that traditional metrics of poverty measurement simply cannot capture. To begin, poverty is fundamentally spatial. Where is poverty located? Within which type of spatial area? Urban centres? Suburban peripheries? Rural hinterlands? Or perhaps in the peri-urban transition zones? Each of these spatial contexts presents unique risk, protective, and policy intervention combinations. Second, poverty is multidimensional. This is a simple way of saying it encompasses insufficient income, poor housing, and lack of access to health and education, inadequate transport, poor environment, lack of social capital, and other factors, leading to a complex non-linear interactive web of poverty. Third, the available data is of poor quality, that is, both are inadequate, survey responses are vague, and administrative data can be incomplete. Furthermore, the poverty measures lack clear boundaries, and many people struggle to classify themselves in terms of poverty, near-poverty, or non-poverty.

Although traditional statistical and econometric approaches are good at estimating poverty and identifying correlates and testing the impact of policies, they have clear rigid assumed boundaries that can obscure these complexities. The continuum of vulnerability is lost in binary classifications. The presence of a spatial boundary is often a poor representation of truncation, overlap, and fuzzy gradation in spatial areas. Imprecise self-reporting, vague definitions of housing quality, and subjective measures of community well-being are all reasons to reject deterministic modelling.

The chapter introduces fuzzy soft sets as a new mathematical construct that addresses the mentioned gaps in the analytical toolbox available to poverty scholars and practitioners. Fuzzy soft sets originate from the fusion of fuzzy logic and soft set theory. Fuzzy logic deals with membership and vagueness through flexible membership functions instead of yes/no dichotomies. In contrast, the soft set theory provides a parameterised mechanism to address multiple criteria without having to assign weight a priori to the criteria. Such a fusion allows analysts to:

- Understand and model poverty along a continuum, rather than as a set of discrete categories with arbitrary cut-offs, recognising, for example, that households can range from extreme deprivation to marginal vulnerability and to relative security
- Address multiple poverty dimensions that are distinct and that may pertain to different parameters: income, housing, health, education, transport and so on
- Continually manage data imperfection and incompleteness, including absent data, subjectivity, and gaps in assessment and measurement
- Create detailed poverty maps that show not only areas with high concentrations of poverty, but also transitional zones, poverty pockets in affluent areas, and areas with particular patterns of vulnerability
- Facilitate multi-criteria decision support for resource targeting to determine which interventions ought to be prioritised based on a composite vulnerability index and describe the trade-offs made

The last two decades have seen a robust development of the theoretical underpinnings of fuzzy soft sets in the mathematical literature, and the emergence of applications in areas as varied as medical diagnosis, environmental risk, supplier selection, and pattern recognition, to name a few. The gap, in this case, is their value for poverty analysis, especially spatial poverty mapping and the design of policy responses. This chapter attempts to fill this gap by showing how fuzzy soft set techniques can be applied in the context of working poverty research.

With the aid of carefully crafted case illustrations, we show how fuzzy soft sets aid in the identification of overlapping poverty clusters that other methods may be unable to identify, the ranking of districts by composite vulnerability for the purpose of focused strategic interventions, and the development of spatial dashboards that describe poverty landscapes in a mathematically rigorous way and, in the case of non-technical users, a way that is easy to understand. All of these case illustrations point to the framework's capacity to deepen and broaden the understanding of the various dimensions of spatial poverty, especially when it comes to defining the differences in the natures of poverty in the various districts, e.g. the difference between districts that face a lack of decent work and those that face unaffordable housing or a lack of access to essential health services.

In addition to methodological innovation, this chapter makes an argument for interdisciplinary cross-pollination that should be more widely appreciated in the field of poverty research. By using mathematical frameworks alongside social science theories of poverty, spatial inequality, and policy, we show, and hope to inspire, the use of formal methods that enhance rather than displace qualitative methods. The fuzzy soft set approach does not supplant ethnographic and advocacy research, participatory policy design, and advocacy, and research, policy, and advocacy. It is an ethnographic and participatory policy design and advocacy. It is a research and policy advocacy approach that is flexible enough to be an analytical scaffolding for cross policy design and advocacy.

It is a policy advocacy approach that is flexible enough to cross various domains of design. It is a fuzzy soft set approach that is flexible enough to cross various domains of design. It is a fuzzy soft set approach that is flexible enough to cross various domains of design. It is a fuzzy soft set approach that is flexible enough to cross various domains of design.

Chapter Overview

To begin with, we start with the conceptual building blocks of fuzzy soft sets and pinpoint the relevant characteristics and operations associated with them and present examples from measuring poverty. We then design a systematic methodology to spatially analyse poverty using fuzzy soft sets wherein practical issues pertaining to the construction of a membership function, parameter setting, and aggregation are discussed. The focal point of the chapter is to present in detail several case studies which help illustrate the extent to which the framework can be used to support policy – from the identification of priority areas for the provision of affordable housing to the targeted distribution of healthcare services to the economically disadvantaged. We end the chapter by taking stock of the road ahead in the field, especially the combination of fuzzy soft sets with machine learning for predictive poverty modelling and GIS for dynamic spatial dashboards.

As more and more data becomes accessible and the challenges of understanding poverty continue to worsen, the fuzzy soft set framework provides an example of a method to address problems where uncertainty is not concealed, where multidimensionality is valued rather than collapsed into one, and where actionable policy ideas are combined with robust analysis. For scholars, the framework creates additional potential for the study of the dynamics of spatial poverty. For policymakers, it improves decision-making support systems. For communities with working poverty, it is more accurately focused and contextually suitable. With this combination of spatial analysis, fuzzy logic, and the principles of policy-oriented research, we hope to advance the theory of poverty and improve the practice of anti-poverty work.

LITERATURE SURVEY

Conceptualising and Measuring Working Poverty

With existing welfare state frameworks failing to mitigate in-work deprivation, working poverty has become an increasingly popular area of study. Lohmann and Marx (2018) offer an in-depth study of working poverty in various European countries and how institutional arrangements, such as minimum wage regulations, tax credit provisions, and social transfer systems, shape the nexus of employment and poverty. This geography of working poverty study found that working poverty exists at various scales, even within the same country, contesting the efficacy of homogeneous policies.

Brady et al. (2021) build on existing theoretical frameworks by focusing on the structural components of working poverty and the role of labour market dualisation, occupational segregation, and household composition within the affluent democracies. In doing so, they add to the structural components of working poverty, as defined by wage inadequacy, job insecurity, insufficient hours, and the lack of employment-related benefits. This also reflects the touch of a fuzzy soft set approach, which is the phenomenon of working poverty, where multiple components can be considered.

Incorporating the multidimensional poverty framework that originated with Sen and later refined by Alkire and Foster, Santos and Villatoro (2018) devise the first operational measures for overlapping deprivations in health, education, and living standards simultaneously. Santos and Villatoro sidestep the most significant alignment with the soft set theory's parameterised approach by recognising that poverty does not result from the absence of a single measure, but rather, a combination of disadvantages. Their methodology does, however, capture a combination of disadvantages from various dimensions of poverty. Yet, the use of crisp cutoffs for deprivation dimensions constitutes the very limitation that soft, fuzzy sets address.

The Spatial Dimensions of Poverty and Inequality

The use of Geographic Information Systems (GIS) and spatial econometric techniques in combination has significantly transformed the analysis of spatial dimensions of poverty. Anselin (2019) reviews the use of spatial regression techniques in the context of poverty mapping and identifies the particular importance of incorporating spatial autocorrelation and spatial heterogeneity in the poverty mapping process. Anselin is the first to demonstrate that the distribution of poverty is not random. Poverty tends to be clustered for various reasons, and these reasons are often due to the infrastructure of a country, historical inequalities, and the process of agglomeration.

Baud et al. (2019) discuss the mapping of poverty in the cities of the Global South, stating that the rapid growth of cities and the different patterns of deprivation are difficult to capture with traditional mapping techniques. They describe the difficulty of setting poverty boundaries in the peri-urban zone where formal and informal settlements and fuzzy boundaries coexist. This condition is perfect for fuzzy spatial analysis.

Gallegati et al. (2020) apply network theory to spatial poverty, and how the pattern of connections (transport, communication, and social) influences the distribution of economic opportunities and the associated risks. This analysis suggests that poverty maps should incorporate not only the area characteristics but also the position of the area in the overall network, which is yet another dimension of complexity that can be captured through fuzzy soft sets with appropriate parameters.

Weeks et al. (2020) enhance the methodologies for the detection of poverty hotspots using data from remote sensing and data from household surveys. The combination of various data sources with different spatial granularity causes data uncertainty in the classification of poverty, and the author uses techniques from probability to address this uncertainty. Using fuzzy soft sets is another way to deal with spatial uncertainty that results from the combination of various sources of data while keeping intact the individuality of each type of data.

Fuzzy Set Theory and its Use Cases

Fuzzy set theory was first described by Zadeh (1965), and since then, it has been widely used in situations where something is vague and unclear. One of the first to introduce a fuzzy method in measuring poverty was Cerioli and Zani (1990). They proposed that the movement from a non-poverty to poverty state was gradual, not sudden. Their work opened the frontiers for the fuzzy poverty indices domain.

Betti et al. (2018) propose fuzzy measures of multidimensional poverty where one can partially be a member of a poverty state. In their model, a given individual or household can be placed in more than one category of poverty at the same time and to different extents depending on the level of deprivation. This model has been used to analyse the statistics of the European Union and is found to be better than the headcount measures in identifying the at-risk population.

With the introduction of the totally fuzzy and relative approach to poverty measurement, Cheli and Lemmi (1995) has been widely used when measuring poverty in Europe. In their model, the headcount was constructed from the sets of deprivation indicators without the use of a threshold. Betti and Verma (2019) recent improvements add more complexity by introducing time and can measure changes in poverty over time.

D'Ambrosio and Chakravarty (2019) discuss the theoretical foundations of fuzzy poverty metrics, focusing on the axiom system and how it enables poverty measurement comparisons across different groups and time periods. Their work shows conditions that fuzzy poverty metrics can be mathematically constructed that will be consistent with desirable metrics properties, including the metrics being monotonic, transfer sensitive, and decomposable with respect to subpopulations.

Soft Set Theory and Extensions

Molodtsov (1999) first proposed soft set theory, which presented a new way of handling uncertainty that is different from both probability and fuzzy logic. One of the most important features of soft sets is that they can handle several different parameters without having to use a weight function, a probability distribution, or a set of probabilities. Maji et al. (2003) constructed the first operational algebra of soft sets, providing the first systems of soft sets for which the operations of union, intersection, and complement could be defined.

With the introduction of soft matrices, Çağman and Enginoglu (2010) provided the first computational framework for the operational algebra of soft sets. Their development of soft matrices for the operational algebra of soft sets allows one to design simple, yet effective algorithms. Their soft matrices have been useful for operationalising frameworks that facilitate decision making by providing several methods for the comparison of options from multiple criteria. This is especially important with large data sets in which the spatial data contain thousands of geographic entities.

Feng et al. (2010) design soft discernibility matrices that are adjustable. The authors demonstrate that soft sets assist multi-attribute group decision analysis which is particularly applicable within poverty policy where multiple contested stakeholders' decision-making resource allocation are involved.

Fuzzy Soft Sets: Integration and Applications

The merged use of soft set theory and fuzzy logic has proven to be useful and applicable in many dimensions of research. The concept of fuzzy soft sets was first introduced by Maji et al. (2001) who defined its basic operations and studied its various properties. Most importantly, the authors illustrated that fuzzy soft sets are able to deal with the vagueness of the fuzzy sets and the parameterisation of soft sets, which provided a more versatile tool to work with in the decision-making processes.

In his work, Ahmad and Kharal (2009) built mappings on fuzzy soft classes. These mappings were useful in extending the theory and provided the potential for complex analytical operations. In the work of Roy and Maji (2007), it was proposed that fuzzy soft set theory be generalised to decision-making situations. The authors are amongst the first to demonstrate the application of fuzzy soft set theory in personnel selection and medical diagnosis. The algorithm proposed by the authors is designed to be used in the aggregation of information provided by fuzzy soft sets in order to rank the alternatives within a set. This is particularly useful in a practical situation, such as prioritising intervention in poverty zones.

Feng et al. (2011) suggested new functions for fuzzy soft sets, which include the operations of AND, OR, and complement that maintain the fuzzy and soft characteristics after such modifications. The new operations enable better exploration of the relationships among parameters. For example, they may help answer the question of how housing inadequacy AND healthcare inaccessibility jointly contribute to vulnerability.

Peng and Garg (2018) create interval-valued fuzzy soft sets to deal with cases where even the degree of membership cannot be determined but characterised by an interval. This extension is important for poverty situations, as the measurement uncertainty aggravates the already vague limits of poverty.

Decision Making With Fuzzy Soft Sets

The use of fuzzy soft sets in multi-criteria decision-making is growing. Çağman and Enginoğlu (2011) provide algorithms for fuzzy soft set based decision making and provide systematic methods for identifying optimal choices from feasible sets. In district-level composite vulnerability ranking, their choice value method, which derives summary scores from fuzzy soft set data, is likely usable for ranking scores.

Kong et al. (2014) develop methods for achieving a consensus in group decision making and apply fuzzy soft sets in situations where multiple decision makers or stakeholders need to resolve conflicting views. In the context of poverty policy, this approach may support participatory prioritisation involving state officials, representatives of the private sector, and community activists.

Han and Wei (2011) created for fuzzy soft sets distance and similarity measures which allow for a quantitative analysis of poverty profiles of various regions, and of different periods in time. These instruments are useful for conducting cluster analysis to identify areas with homogeneous poverty characteristics, thereby identifying categories that may be relevant for the design of policies.

Qin et al. (2020) combine the outranking method in multi-criteria decision analysis with fuzzy soft set methods, developing hybrid approaches that utilise the advantages of both. For instance, Qin et al.'s ELECTRE-based fuzzy soft set method offers robust rankings even when preferences among the criteria remain unspecified, which is often the case in poverty policies because stakeholders may have unclear or conflicting priorities.

The Use of Spatial Analysis and GIS

The combination of fuzzy methods and Geographic Information Systems offers a number of opportunities in spatial decision support. For example, Ghadiri et al. (2018) created a fuzzy GIS approach for the environmental vulnerability mapping and showed how fuzzy membership functions can be used to model spatially varying risk zone boundaries. The environmental indicators that Ghadiri et al. used to build spatially varying fuzzy membership surfaces can be used in a similar manner for poverty mapping.

Aburas et al. (2019) used land use planning decision processes and fuzzy soft set theory to demonstrate how the framework can integrate stakeholder preferences, and the spatial framework of the problem. His work illustrates how fuzzy soft sets can be used within GIS to develop more sophisticated decision support systems and spatial dashboards for poverty.

Akbari et al. (2020) adjust fuzzy soft sets and spatial multi-criteria evaluation with basic GIS overlay operations. They're doing an example with composite vulnerability indices which is adaptable to determining potential optimal target locations for interventions to reduce poverty, such as job training centres, healthcare clinics, or affordable housing.

Machine Learning and Predictive Modelling

The most recent research attempts to combine soft sets and machine learning. One such is Zhang et al., (2020) who created a classification algorithm based on fuzzy soft sets and it was found to be superior to other crisp based classifiers for a given domain with borderline classes. Such an algorithm, with appropriate modifications, could be used to determine classification of households or spatial units into various levels of poverty (high, medium, low, or not poor) and would take into consideration the slight changes that can occur to move from one level to another.

Ali et al. (2021) used fuzzy soft sets and neural networks for pattern recognition, and used poor sample training data. This integration of neural networks with fuzzy soft sets for pattern recognition yields adjustable and learned models, which is a useful feature for developing models for poverty prediction based on historical data, coupled with measurement errors, and missing or incomplete data.

Jana and Pal (2019) suggest multi-granulation fuzzy soft sets as a way to capture and express different levels of detail or resolution. Such a hierarchical method can be applied to the different levels of poverty measurement at varying spatial scales—from individuals to households to neighbourhoods, districts, and regions—facilitating an analysis that acknowledges the hierarchical (or nested) spatial data structure.

Policy Applications and Decision Support

The use of fuzzy soft set theory is yet to be researched thoroughly in the field of social policy. However, there is some evidence of fuzzy methodologies employed in social policy. Nardo et al. (2019) focused their criticism of composite indicators in the policy-making process on arbitrary weighting and offer, as a more informative way, fuzzy aggregation to customise the retention of mass information on the components. Their remarks justify the use of fuzzy soft sets to avoid pre-weight set assignments.

Atkinson and Marlier (2010) provided the most in-depth, yet, without the use of fuzzy methodologies, examination of the social exclusion and poverty indicators in the European Union. They raised, and offered in detail, most of the problems of indicator construction, including systematic bias of cross-country thresholds and their measurement. These are the problems that fuzzy soft sets would be able to solve.

The discussions of Jolliffe and Prydz (2021) on measuring poverty and the Sustainable Development Goals focused on the need for a fuzzy approach as a result of varying country contexts, differing availability of data, and distinct policy frameworks. These are the problems that fuzzy soft set methodologies would address.

GAPS AND OPPORTUNITIES

As poverty analysis relies spatially on fuzzy soft set theory, it remains under-analysed in spatial poverty analysis literature. Second, fuzzy soft sets and other spatial econometric techniques have yet to be merged, so hybrid approaches remain unexplored. Third, development and operational guidelines for the construction of poverty context-specific membership functions and parameters remain lacking, and thus, require context-specific methodological elaborations.

Although fuzzy soft set theory has been applied in social science, it has remained mainly social-scientific, lacking the mathematical discourse necessary. This chapter demonstrates mathematical rigor in working on poverty applications and closes this gap.

The literature survey uncovers a unique intersection between fuzzy soft set development and empirical ingenuity, with respect to spatial poverty and the lack of intersection between the two. This chapter situates itself at such an intersection, utilising computer science and operations research-inspired mathematical frameworks, responding to the challenges in poverty and spatial analysis. The synthesis unlocks new pathways for innovation that are of empirical policy relevance, coupled with the advancement of theory.

THEORETICAL FOUNDATIONS

THE USE OF FUZZY soft sets in spatial poverty analysis requires consideration of three interconnected branches of mathematics: fuzzy set theory, soft set theory, and the combination of the two. This chapter presents these mathematical branches as the analytical building blocks and illustrates how each one addresses the various challenges associated with conventional methods of measuring poverty.

Fuzzy Set Theory: Dealing with Uncertainty and Degrees

In classical set theory, there is a rule of two. An element is either a member of the set or it is not. In the analysis of poverty, this means a household is either poor or not poor. This will give a household, or an entire country, the opportunity to be on the wrong side of a poverty line. A single household that earns 49% of the poverty line will be in the same category, with all of the associated disadvantages, as a household that earns 51% of the line. This illustrates the fundamental unsuitability of this approach.

Zadeh's fuzzy theory holds that one of the ways to address this problem is to allow for a partial set of memberships. Units have, or do not have, a set membership, or may have, with a membership degree anywhere between 0 and 1. A membership equal to 0 is a complete lack of membership to the set, a membership equal to 1 is a full membership to the set, and an intermediate value reflects a partial membership to the set. In terms of poverty analysis and fuzzy set membership, this is especially true for all households.

Every fuzzy set has a distinct membership function that assigns a membership degree to every set element. Households can be assigned poverty membership levels according to their income-to-poverty-line ratios. A household that makes 40% of the poverty line may be ascribed a membership degree of 1.0 (totally poor), a household that makes 70% may have a degree of membership of 0.6 (moderately poor), and a household that makes 120% may have a degree of membership of 0 (not poor). The membership function can be adjusted and is a reflection of how poverty is perceived and can vary based on policy goals or real-world data.

Like classical sets, fuzzy sets support similar operations. The maximum operator is used for the union of two fuzzy sets. A household that has a membership degree of 0.7 in “Income Poor” and 0.4 in “Health Deprived” will have a membership degree of 0.7 in “Income Poor OR Health Deprived.” For intersection, the minimum operator is used. The same household is assigned a membership degree of 0.4 in “Income Poor AND Health Deprived.” It is thus possible to conduct a multi-dimensional analysis of poverty in which households are impacted by a mix of different factors to different extents.

To define variables linguistically is to apply fuzzy reasoning to fuzzy sets. When setting boundaries for impoverished populations, policymakers use phrases like 'moderate deprivation' or 'vulnerable households,' instead of employing exact metrics. Formalising this, variables are defined for fuzzy sets such that 'Very Low,' 'Low,' 'Medium,' 'High,' and 'Very High' correspond to fuzzy sets through distinct membership functions. Thus, on qualitative determinations from communities or specialists, community feedback can, through mathematical processing, retain qualitative community feedback variables.

Fuzzy and classical sets are connected through an alpha-cut operation, which removes all elements of a set that fall below a specified alpha. Different values of alpha correspond to different subsets of a poverty population. For instance, an alpha of 0.3 gets all at-risk households, while increasing alpha to 0.9 gets only those in extreme poverty. Instead of setting just one arbitrary value, policymakers can analyse poverty data for multiple values of alpha.

Soft Set Theory

Diminished understanding and consideration of dimensions concerning poverty such as its multidimensionality and its various parameters reflects a core issue of the poverty characterisation framework. Not all households face the same dimensions of poverty. A household may have low income, live in poor quality housing, lack access to healthcare, have limited education, or may have other issues. Existing methods for addressing such dimensions tend to look at a single dimension, or they combine multiple dimensions, but how do we choose the weight for income as compared to housing or health is a question to be addressed.

Soft set theory offers an alternative with its approach to parameterisation. A single element of a soft set does not have a membership function, and no element has an associated probability. Instead, it associates each parameter with the subset of elements pertaining to that parameter. Each of the parameters is kept separate and not collapsed into a single attribute, thereby maintaining the dimension and selectively retaining the associated variables.

For example, a soft set over a universe of spatial units (e.g. districts) with respect to poverty parameters is determined by the districts that meet each parameter. If parameters are defined as "unemployment," "inadequate housing," and "poor access to healthcare," the soft set specifies which districts have "inadequate housing," which have "poor access to healthcare," and which have "high unemployment." This is described by a matrix with districts as the rows, parameters as the columns, and a cell equal to 1 if that district exhibits that parameter and 0 if it does not.

Flexibility is one of the reasons this approach works best. There is no need to merge the parameters with fixed weights, and different decision-making criteria can be used depending on the context of the policy. In the case of comprehensive interventions, districts with a larger number of parameters may be selected. In the case of targeted interventions, selection is determined through specific parameter combinations. The soft set retains all information needed for different analyses without having to commit to any single aggregation approach.

Operations of soft sets allow for the combination or comparison of different characteristics of poverty. The union combines parameters from different sets, which is useful when data from different surveys is being pooled. The intersection only retains parameters that are contained in both sets, which is useful for determining consensus dimensions of poverty. The AND operation makes compound parameters, such as "income poverty AND poor healthcare," which helps to identify districts that possess both characteristics.

The choice value method, in this case, is critical for decision-making. Each district's choice value equals the number of parameters that the district possesses. Therefore, districts that have a high choice value, and as a result possess a greater number of poverty characteristics, are more likely to be chosen for intervention. This method also provides weightless multi-criteria decision-making, which means that all parameters are considered equally important, thus, maintaining information about which specific characteristics each district possesses.

Integrated Framework of Fuzzy Soft Sets

The fuzzy soft set model combines the benefits of soft sets and fuzzy sets and addresses the challenges of vagueness and parameterisation at the same time. In the case of fuzzy soft sets, each parameter is not related to a crisp subset, but rather to a fuzzy subset. This means that not all the elements of the subset satisfy the parameter or, rather, they satisfy the parameter to different degrees.

In the case of poverty analysis, we can think of the universe as districts and the parameters as indicators of poverty. For the parameter "high unemployment," rather than classifying each district as a fuzzy soft set, it would assign membership values for each district, for example, district A might have a membership value of 0.9 (very high unemployment), district B might have a membership value of 0.3 (relatively low unemployment), and district C would have a membership value of 0.6 (moderate unemployment). This double flexibility is powerful in parameterisation because it preserves multidimensionality, and in fuzzification because it captures the gradation for each dimension.

The tabular representation is similar to a soft set matrix, but the cell entries are not binary and can range from 0 to 1. Each cell shows the degree to which a certain spatial unit possesses a certain characteristic of poverty. The complete poverty profile of a district is comprised of the membership values it holds with respect to all the parameters, and it reflects the dimensions of poverty that are problematic, as well as the intensity of deprivation along each dimension.

The operations pertaining to fuzzy soft sets extend fuzzy and soft set operations. For union, combining parameters and taking the maximum of the membership degrees is performed. For intersection, the retention of common parameters and the taking of the minimum memberships is done. The AND operation of the soft sets produces compound parameters that reflect the membership equal to the minimum of the component memberships.

The accumulation of memberships of the various parameters is the information aggregation performed by the fuzzy choice value. For example, a district is characterised by income poverty membership of 0.8, housing deprivation of 0.6, and 0.7 in healthcare deprivation. Thus, the fuzzy choice value of the district is $0.8 + 0.6 + 0.7 = 2.1$. This value is the composite vulnerability score which represents the number of dimensions of poverty and the severity. The districts with the highest to lowest fuzzy choice values can determine the order of greatest need to address the problem.

In the instance where parameters differ in their importance, the weighted fuzzy choice value involves the multiplication of each membership by the weight of the parameter and summing the result. This feature is beneficial because, in the absence of weights, the value of the unweighted choice still reveals a lot and the weights can be modified to determine the sensitivity of the analysis to assess the impact of different prioritisation on the order of the values.

The fuzzy soft set framework provides unique means of tackling long-standing challenges in poverty measurement. First, it manages, simultaneously, both vagueness and parameterisation, closing a fundamental gap. Fuzzy sets capture gradual changes, but also require weighted aggregation of dimensions. Soft sets maintain multidimensionality without imposition of weights, but do have crisp boundary stipulations. Fuzzy soft sets tackle fuzzy boundaries on all dimensions and maintain separation of dimensions.

Second, the framework's characteristics fit perfectly with the imprecision of available data. In the case of real data on poverty, the issues of measurement error, missing data, subjective value assignments, and temporal misalignment are all present. Fuzzy membership degrees, in this case, can reflect the uncertainty of measurement. Missing data and measurement errors can be modelled in different ways, such as reducing the number of dimensions in the parameter space for certain affected units. The absence of data due to the subjective stigmatization of poverty corresponds to the fuzzy value without requiring an imprecise value attribution.

Third, fuzzy soft sets enable and support analysis without thresholds, while also providing the means to examine sensitivity to thresholds. The framework provides an alternative to ex-ante determination of a poverty line by first ranking units in space (geographically) according to the value of the variable of interest selected by the analyst. This allows the policy-maker to specify where a funding line would be in order to determine the number of districts prioritised for assistance. In this way, the alpha-level sets illustrate the changes in poverty owing to variations in the severity of the threshold (or in the case of the absence of a threshold).

Fourth, a system that preserves the richness of the information that supports different policy needs. One representation supports more than one analysis: comprehensive rankings through choice values, pinpointing certain profiles of a district through parameter filtering, cluster analysis of the different types of poverty, and analysis of trends through different intervals of time. This is in contrast to approaches that streamline data into singular indices.

Fifth, fuzzy soft sets offer decision support that is interpretable and transparent. Unlike "black box" machine learning approaches, the processes are explicit and can be traced. For any district's ranking, one can evaluate its degree of membership in each parameter and understand velocity in that position. Such transparency is important in communicating with policymakers and the public.

The framework also easily incorporates multi-scale analysis and spatial heterogeneity. Different regions may experience different types of poverty, each requiring different parameters. Urban poverty, for instance, can be housing affordability, while in the rural context, it can be about the viability of agriculture. Through the use of aggregation operators, multi-scale analysis can be achieved. For example, fuzzy, soft sets at the household level can aggregate to the neighbourhood level through averaging membership scores.

These theoretical advantages justify fuzzy soft sets being viewed as a framework that, while providing analytical rigour, also preserves the complexity of poverty. It offers rich information and supports decision-making, preserves mathematical rigour, and remains interpretable to different stakeholders. The following sections deliver practical spatial poverty analysis and policy design, showcasing these advantages.

CASE STUDY: IDENTIFYING OVERLAPPING POVERTY CLUSTERS

This study analyzes the spacial distribution of the phenomenon of poverty across the units of the study, comprising twenty-five administrative divisions of a single state of the study. The state of the study displays notable extremes in economic, geographic, and demographic attributes. This state-of-the-art exhibits multiplicities and complexities in the manifestation of poverty. The northernmost districts or units represent former industrial zones where the collapse of manufacturing industries led to a situation of concentrated unemployment and to the phenomenon of market labour distress and the collapse of public services. However, these units still enjoy a plentiful and relatively adequate housing stock and public service infrastructure from their former economic prosperity. The central peri-urban belt which encompasses the state capital has rapid economic growth which is driven by rural-urban migration. The southernmost districts enjoy a tourism-dependent economy which has created seasonal employment and relatively high average incomes, although these conditions are not equitably distributed across the districts.

DATA COLLECTION

The various data sources used in this case study illustrate the excellent integration of administrative records required for a thorough assessment of different dimensions of poverty. Integrated administrative records and multiple surveys from a sample of 500 households per district (12,500 statewide) provide data on employment status, income, household structure, and self-reported deprivation. Records from the Department of Labour provide district-level data on unemployment, labour force participation, and employment by economic sector. Data on the quality of the housing stock and housing overcrowding, water and sanitation accessibility, basic amenities, and housing cost to income ratio comes from the housing authorities. From the records of the Department of Health, we obtain data on the density of healthcare facilities, the ratio of healthcare providers to the population, the prevalence of diseases, and the rates of utilisation of healthcare services. The educational data available include school enrollment rates, the ratio of students to teachers, educational infrastructure, quality indices, levels of educational attainment, and rates of utilisation of educational facilities.

The administrative records and the various surveys used to assess the different dimensions of poverty reflect a different level of spatial, temporal, and measurement resolution. While they provide household-level detail, surveys produce sampling uncertainty. Administrative records are comprehensive, but the temporal lag and gaps in reporting are common.

The different levels of quality demonstrated by the available records, from low to high, are not a limitation for the fuzzy soft set framework, but rather a strength.

The data preparation process included a number of steps. First, all district-level indicators were normalised to the same scale, typically 0 to 1, to allow for comparison across indicators. Second, within each poverty dimension, indicators were combined to provide each dimension a score. For the dimension of employment poverty, a composite score of employment deprivation was created which combined the rates of unemployment, informal employment, and income inadequacy. For housing poverty, the deprivation scores were aggregated including rates of overcrowding, amenity inadequacy, structural quality deficiency, and affordability stress. For healthcare poverty, the integration included the diseased burden, underutilisation of services, and the scarcity of facilities and providers. Education poverty deprivation scores were combined to include gaps in enrolment, deficits in infrastructure, and shortages in teachers and attainment.

Third, membership functions were developed for each dimension of poverty to convert normalised scores to fuzzy membership degrees. A sigmoid membership function was selected for employment poverty to capture the phenomenon of employment poverty aggravating almost exponentially when there is lack of employment; an example of this is when there is moderate unemployment there is an enduring hardship, whereas in severe unemployment there is catastrophic deprivation. Calibrations were made to indicate that districts with employment deprivation scores lower than 0.2 would receive a membership lower than 0.3 (low poverty); scores between 0.2 to 0.6 would have memberships that linearly increased from 0.3 to 0.7 (moderate poverty); and scores that were greater than 0.6 would receive memberships greater than 0.7 and at most 1.0 (severe poverty). For poverty in housing, healthcare, and education, similar membership functions were created and tailored to the specific characteristics of each dimension.

Fuzzy Soft Set Formulation

The fuzzy soft set for this case study has a universe U comprising the twenty-five districts labelled D_1 through D_{25}. The parameter set A includes the four basic dimensions of poverty: employment poverty (e_1), housing poverty (e_2), healthcare poverty (e_3), and education poverty (e_4). The fuzzy soft set (F, A) attributes to each parameter a fuzzy set over the districts, where the degrees of membership indicate the levels of poverty in the respective districts.

The whole fuzzy soft set can be described in terms of a matrix M of dimensions twenty-five by four, where each row stands for a district, and each column represents a parameter of poverty. An entry m_ij captures the degree of membership of district i to poverty parameter j, which can translate to zero (absence of deprivation in that dimension) and one (complete deprivation). For example, considered entries would include: District D_1 (one urban industrial centre) with memberships [0.85, 0.45, 0.30, 0.25], which illustrates a case of extreme poverty in employment, a case of moderate housing problem, and a case of relatively sufficient health and education. District D_7 (one peri-urban area of informal settlement) with memberships [0.55, 0.90, 0.70, 0.60], which illustrates a case of moderate poverty in employment but extreme deprivation in housing, and also some serious deficiencies in health and education. District D_{15} (one of the remote rural districts) with memberships [0.40, 0.35, 0.85, 0.75], which illustrates a case of moderate employment and housing problem, but extreme deficiency of health and education due to the rural area and service delivery problem.

This matrix representation allows for a complete description of each district individually with respect to its multidimensional poverty profile and also allows for the necessary computations to be done so that the district can be allocated to a specific cluster. The fuzzy soft set also captures the fact that poverty is multidimensional, and that it is often the case that some dimensions are cut off for some districts, meaning that some districts are very poor on one dimension but only moderately poor on others, and others are moderately poor in several dimensions at the same time.

Figure 1. Poverty and policy solutions in districts

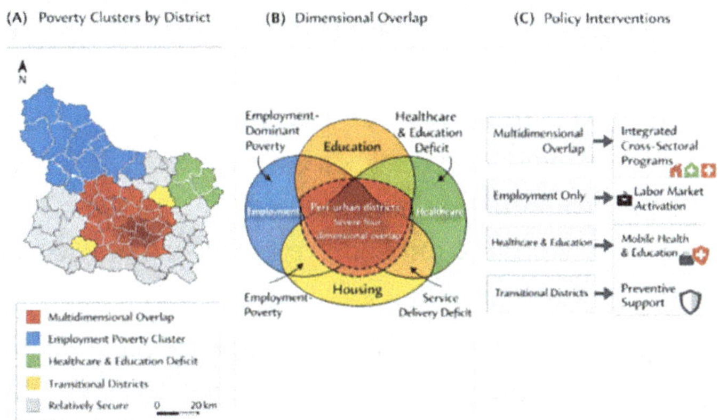

Cluster Identification

Determining overlapping poverty clusters for fuzzy soft sets is a description and involves a methodical approach. First, for each employment, housing, health care, and education poverty parameter, membership degrees for each dimension are analysed and fuzzy clustering techniques are applied to create clusters. For poverty parameter e_1, employment, districts are divided by membership score: severely affected are those above 0.7, moderately affected are those between 0.4 and 0.7, and low membership is below 0.4. Districts close to the threshold, though, may display partial memberships, so classification is a little fuzzy and logic districts may fall into. The same approach is applied to the clustering for housing, health care, and education poverty separately. Each dimension shows a different, distinct spatial pattern.

High membership districts that are spatially close to each other are spatially high membership districts, and the distance clustering within and between high membership districts is measured using spatial autocorrelation statistics. The next step is overlapping clusters, which involves the analysis of the other dimensions. The other parameters are combined using an AND operator so that the dimensions of poverty (e.g. employment and housing) are minimally defined by their membership scores.

The above excerpt reflects the aggregation of the most extreme clusters of vulnerability, which in this case, are obtained from fuzzy choice values, which are calculated by summing the memberships across all four poverty parameters. Districts with a fuzzy choice value of greater than three reflect the most extreme level of multidimensional poverty whereas, a fuzzy choice value averaging between two and three indicates moderate levels of poverty. From this, it could be seen that the survey captures the level of poverty in a more comprehensive sense, as opposed to a singular view from a single dimension.

In the end, poverty cluster overlap analysis demarcates the linkages between dimension-specific clusters and comprehensive clusters, with the aim of identifying districts that fall into more than one category of extreme poverty and to overlap the clusters within the context of a Venn diagram. This serves the purpose of identifying the extent to which the poverty dimensions operate in isolation versus the extent to which they act in tandem, and in the latter case, it points to the existence of more complex poverty conditions that call for comprehensive and coordinated solutions.

RESULTS AND SPATIAL PATTERNS

The analysis identifies overlapping patterns of poverty clusters with specific implications for policy. The most severe poverty cluster due to lack of employment includes seven districts, most of which are former industrial centres in the north of the country, where the manufacturing base is shrinking. These districts have employment poverty membership rates from 0.75 to 0.92, with varying membership across the other dimensions of poverty. Three of the districts (D_1, D_2, D_3) have employment poverty and little to no membership in the other dimensions of poverty, such as housing, health, and education, which are all below 0.5. These districts have, for their former industrial centres, 'good' legacy infrastructure and service provision, although employment levels are low. The policy implication is that labour market focused interventions are likely to be needed. This includes interventions such as job training and economic diversification, as well as support for the unemployed, which may address the most important poverty driver, without a need for extensive multi-sectoral shifts.

The concentric zone model comprises nine districts focusing on the emerging peri-urban zones that are rapid informal settlement clusters. Housing poverty memberships exceed 0.75, indicating deficiency in basic amenities, structural quality, overcrowding, and affordability stress. Despite this, the overlap analysis indicates two sub-patterns. Five districts (D_6, D_7, D_8, D_9, D_{10}) combine extreme housing poverty with employment poverty (moderate to severe D_6, D_7, D_8 0.50 to 0.70; D_9 D_{10} 0.70 to 0.80), healthcare poverty (D_6, D_7, D_8 0.60 to 0.80), and education poverty (D_6, D_7, D_8 0.50 to 0.70). Their fuzzy sets choice values (2.7 to 3.4) show multidimensional poverty. This indicates that the housing deficit should be addressed with integrated solutions on employment, health, and education. In the case of the three (D_{11}, D_{12}, D_{13}) districts, extreme housing poverty and low memberships on the other dimensions imply that single-focus, housing poverty interventions are likely to be efficient.

The healthcare poverty cluster has ten districts. Most are rural and remote and are characterised by geographic isolation and low population density, which significantly hinders service delivery. Healthcare memberships range from 0.70 to 0.95. Overlap patterns differ considerably. Four districts (D_{14}, D_{15}, D_{16}, D_{17}) merge healthcare poverty with severe education poverty (0.70 to 0.85 memberships), highlighting service provision shortcomings in these remote locations. Their moderate employment and housing poverty memberships (0.35 to 0.50) indicate that economic conditions are less problematic than access to services. Policies should prioritise mobile health services, telemedicine, and teaching accommodation. Three districts (D_{18}, D_{19}, D_{20}) show healthcare poverty with severe housing poverty, but moderate employment and education poverty, which may be attributed to informal settlements close to urban areas that are poorly served by health facilities.

The education poverty cluster has eight districts with memberships greater than 0.65. Overlap analysis indicates education poverty often coincides with healthcare poverty—six of the eight education-poor districts have healthcare memberships greater than 0.60. This points to shared obstructing factors: geographic remoteness, gaps in service delivery systems, and deficiencies in the recruitment of healthcare professionals.

Four districts with overlaps in healthcare, education, and housing poverty (D_7, D_{15}, D_{16}, D_{21}) exhibit compounded vulnerability that calls for more integrated approaches.

The northern industrial region reveals a geographic concentration of overlapping poverty, particularly employment poverty, with moderate overlap to other dimensions. Employment poverty, combined with other acute forms of poverty (housing, healthcare, and education), characterises the peri-urban belt around the state capital—a concentration of seven districts with fuzzy choice values greater than 2.8. The remoteness of the eastern mountain region is marked by a pronounced overlap of healthcare and education poverty, with moderate overlap on other dimensions. Similarly, the southern coastal region is marked by sparse poverty across the dimensions, with maximum memberships rarely rising above 0.5.

The first critical overlap zone comprises five districts, specifically D_7, D_8, D_{15}, D_{16}, and D_{21}, with fuzzy choice values between 3.0 and 3.4. Here, districts include membership above 0.6 in at least three poverty dimensions. These zones are critical, as they require immediate multi-sectoral interventions in employment, housing, health care, and education. Moderate overlap zones comprise eight districts, two of which have membership above 0.5 in two dimensions. This indicates the need for dual-focus interventions, while in low overlap zones, which comprise seven districts and where the districts are in severe poverty in at most one dimension, these are amenable to targeted sectoral programmes.

Figure 2. Cluster analysis and poverty interventions

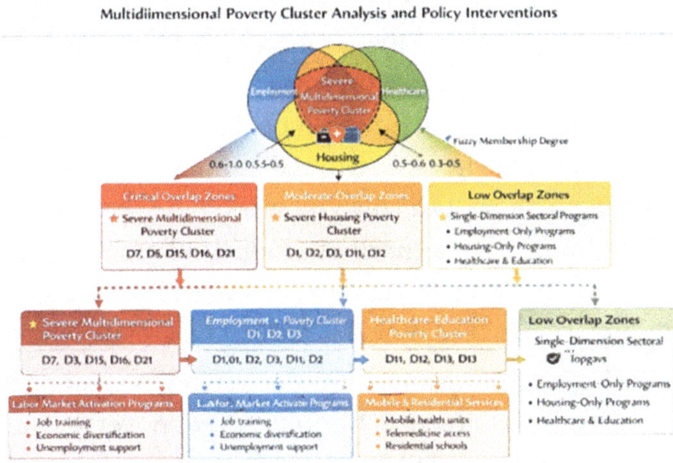

POLICY IMPLICATIONS

Based on the overlapping cluster analysis, integrated poverty reduction programmes were recommended for each of the five districts with overlapping severe poverty due to the policy recommendations being described for each of the various aspects of poverty. The programmes would need to be comprehensive and integrated in order to be effective, combining employment programmes, such as job training and entrepreneurship support, with housing improvements to include slum upgrading and the construction of affordable housing, expansion of healthcare to include mobile clinics, and education improvements which include infrastructure improvements and scholarship programmes. When partnering various components as described above, the effectiveness of the intervention is reinforced, as enhancements in one element aid other components.

For districts with severe and only one or two dimensions of poverty, combined with low overlap, more targeted sectoral interventions are suggested. For example, in the districts burdened with employment poverty, focus should be on labour market programmes and not on housing or service expansions. Healthcare-poverty ‘islands’ would require tailored interventions of healthcare services, particularly telemedicine.

The analysis also identified the need for synergies of cooperation due to the existence of certain spatial attributes. The state capital's poverty belt that is peri-urban will require metropolitan planning that encompasses the integration of urban growth with affordable housing, and the peri-urban transport network. The analysis also identified the spatial concentration of employment poverty in the northern industrial area which indicates a need for integrated planning to support regional economic growth so that employment opportunities and industries can be created and attracted.

Districts with moderate poverty present a window of opportunities for preventative measures to stop the slide into severe poverty. These districts can be endowed with upgraded social protection systems, early poverty indicator warning systems, and rapid response mechanisms to deal with the emerging issues.

There needs to be a greater refinement of resource allocation in the targeting of anti-poverty expenditures to move from a more simplistic approach of arbitrary geographic equal allocation, one-dimensional poverty line-based allocations, or single poverty index allocation. There can be fuzzy choice values allocations, whereby districts are funded based on the various dimensions and indicators of poverty they possess. Also, funding streams can be earmarked to address particular issues in order to meet specific community needs.

The study illustrates that fuzzy soft sets capture overlapping poverty clusters, provide a multi-faceted and geo-spatial understanding of poverty, and assist in the designing and tailoring of policies to specific poverty conditions on the ground rather than the use of overly general solutions.

CONCLUSION AND FUTURE SCOPE

This chapter outlines fuzzy soft sets as an innovative tool for spatial poverty analysis. In particular, fuzzy soft sets are an innovative tool for spatial analysis of working poverty and its complexities, multidimensionality, and uncertainties. Establishing a fuzzy soft set theory framework enables the retention of valuable data often disregarded in previous frameworks. A case study describing the identification of overlapping poverty clusters is used to demonstrate how the framework enables differentiated targeted interventions by identifying and differentiating districts that require a broad, integrated multisectoral approach from those that need specific focused programmatic interventions.

Some of thc positivc attributes of the framework are the definitional clarity in the decision-making process, the ability to analyse and assess various dimensions of a problem, and the absence of a specified threshold for analysis or decision-making based on the resources available to users of the framework. The framework is also able to accept the imperfections that exist in poverty data. The framework also displays a level of spatial intelligence, making the connection between data on poverty and geography. The framework supports the design of targeted, locally tailored strategies for poverty reduction, as opposed to generalised, holistic approaches that often miss the essence of the problem.

Interdisciplinary engagement from a mathematically grounded and social science perspective is also fostered by fuzzy soft sets, and it invokes a rich qualitative discourse that is not displaced. The framework presents a complex view of poverty and provides users with simple analytical frameworks that are appropriate for non-technical users while allowing advanced spatial analysis to take place.

Future research may include the integration of fuzzy soft sets and machine learning for predictive modelling of poverty, the potential of dynamic fuzzy soft sets in longitudinal analyses of poverty, breadth expansion to include climate vulnerability and other impact dimensions, sophisticated contextual adaptations of flexible methodologies to poverty, the use of community participatory approaches to define parameters, and the development of accessible software tools for diverse applications. Such applications aim to transform theoretical advances into practical impacts in the fight against poverty on a global scale.

REFERENCES

Aburas, M. M., Ho, Y. M., Ramli, M. F., & Ash'aari, Z. H. (2019). Improving the capability of an integrated CA-Markov model to simulate spatio-temporal urban growth trends using an Analytical Hierarchy Process and Frequency Ratio. *International Journal of Applied Earth Observation and Geoinformation*, *59*, 65–78. DOI: 10.1016/j.jag.2017.03.006

Ahmad, B., & Kharal, A. (2009). On fuzzy soft sets. *Advances in Fuzzy Systems*, *2009*(1), 586507. Advance online publication. DOI: 10.1155/2009/586507

Akbari, M., Neamatollahi, E., Neamatollahi, P., & Akbari, E. (2020). Prioritizing municipal solid waste landfill site selection using fuzzy-AHP analysis. *Environmental Science and Pollution Research International*, *27*(21), 26283–26294. DOI: 10.1007/s11356-020-08929-y

Ali, G., Alolaiyan, H., Pamučar, D., Asif, M., & Lateef, N. (2021). A novel MADM framework under q-rung orthopair fuzzy bipolar soft sets. *International Journal of Fuzzy Systems*, *23*(7), 2163–2182. DOI: 10.1007/s40815-021-01095-7

Anselin, L. (2019). A local indicator of multivariate spatial association: Extending Geary's c. *Geographical Analysis*, *51*(2), 133–150. DOI: 10.1111/gean.12164

Atkinson, A. B., & Marlier, E. (2010). *Income and living conditions in Europe*. Publications Office of the European Union., DOI: 10.2785/53320

Baud, I., Sridharan, N., & Pfeffer, K. (2019). Mapping urban poverty for local governance in an Indian mega-city: The case of Delhi. *Urban Studies (Edinburgh, Scotland)*, *56*(5), 934–951. DOI: 10.1177/0042098007087341

Betti, G., Gagliardi, F., Lemmi, A., & Verma, V. (2018). Comparative measures of multidimensional deprivation in the European Union. *Empirical Economics*, *55*(3), 1071–1100. DOI: 10.1007/s00181-014-0904-9

Betti, G., & Verma, V. (2019). Fuzzy measures of longitudinal poverty in a comparative perspective. *Social Indicators Research*, *142*(1), 249–272. DOI: 10.1007/s11205-018-1911-5

Brady, D., Blome, A., & Kleider, H. (2021). How politics and institutions shape poverty and inequality. In Brady, D., & Burton, L. M. (Eds.), *The Oxford handbook of the social science of poverty* (pp. 555–578). Oxford University Press., DOI: 10.1093/oxfordhb/9780199914050.013.25

Çağman, N., & Enginoğlu, S. (2010). Soft matrix theory and its decision making. *Computers & Mathematics with Applications (Oxford, England)*, *59*(10), 3308–3314. DOI: 10.1016/j.camwa.2010.03.015

Çağman, N., & Enginoğlu, S. (2011). Fuzzy soft set theory and its applications. *Iranian Journal of Fuzzy Systems*, *8*(3), 137–147. DOI: 10.22111/ijfs.2011.292

Cerioli, A., & Zani, S. (1990). A fuzzy approach to the measurement of poverty. In Dagum, C., & Zenga, M. (Eds.), *Income and wealth distribution, inequality and poverty* (pp. 272–284). Springer., DOI: 10.1007/978-3-642-84250-4_18

Cheli, B., & Lemmi, A. (1995). A totally fuzzy and relative approach to the multidimensional analysis of poverty. *Economic Notes*, *24*(1), 115–134.

D'Ambrosio, C., & Chakravarty, S. R. (2019). Polarization orderings of poverty and richness. *Review of Income and Wealth*, *65*(4), 794–820. DOI: 10.1111/roiw.12385

Feng, F., Jun, Y. B., Liu, X., & Li, L. (2010). An adjustable approach to fuzzy soft set based decision making. *Journal of Computational and Applied Mathematics*, *234*(1), 10–20. DOI: 10.1016/j.cam.2009.11.055

Feng, F., Li, Y., & Leoreanu-Fotea, V. (2011). Application of level soft sets in decision making based on interval-valued fuzzy soft sets. *Computers & Mathematics with Applications (Oxford, England)*, *60*(6), 1756–1767. DOI: 10.1016/j.camwa.2010.07.006

Gallegati, M., Palestrini, A., Petrini, M., & Russo, A. (2020). An agent-based model with network effects on income inequality. *Journal of Economic Interaction and Coordination*, *15*(4), 863–889. DOI: 10.1007/s11403-020-00295-y

Ghadiri, M., Shabanpour, M., Hoseini, S. M., & Askarizadeh, D. (2018). Evaluation of fuzzy analytic hierarchy process in land suitability analysis for crop cultivation (case study: Takestan-Avaj plain). *Applied Ecology and Environmental Research, 16*(2), 1667–1684. DOI: 10.15666/aeer/1602_16671684

Han, B. H., & Wei, M. (2011). Distance and similarity measures for hesitant fuzzy soft sets. *Journal of Information and Computational Science, 6*(4), 265–274.

Jana, C., & Pal, M. (2019). Multi-criteria decision making process based on some single-valued neutrosophic Dombi power aggregation operators. *Soft Computing, 25*(7), 5055–5072. DOI: 10.1007/s00500-020-05509-z

Jolliffe, D., & Prydz, E. B. (2021). Societal poverty: A relative and relevant measure. *The World Bank Economic Review, 35*(1), 180–206. DOI: 10.1093/wber/lhz018

Kong, Z., Gao, L., & Wang, L. (2014). An efficient decision making approach in incomplete soft set. *Applied Mathematical Modelling, 38*(7-8), 2141–2150. DOI: 10.1016/j.apm.2013.10.009

Lohmann, H., & Marx, I. (2018). *Handbook on in-work poverty*. Edward Elgar Publishing., DOI: 10.4337/9781784715632

Maji, P. K., Biswas, R., & Roy, A. R. (2001). Fuzzy soft sets. *Journal of Fuzzy Mathematics, 9*(3), 589–602.

Maji, P. K., Biswas, R., & Roy, A. R. (2003). Soft set theory. *Computers & Mathematics with Applications (Oxford, England), 45*(4-5), 555–562. DOI: 10.1016/S0898-1221(03)00016-6

Molodtsov, D. (1999). Soft set theory—First results. *Computers & Mathematics with Applications (Oxford, England), 37*(4-5), 19–31. DOI: 10.1016/S0898-1221(99)00056-5

Nardo, M., Saisana, M., Saltelli, A., Tarantola, S., Hoffman, A., & Giovannini, E. (2019). *Handbook on constructing composite indicators: Methodology and user guide*. OECD Publishing., DOI: 10.1787/9789264043466-

Peng, X., & Garg, H. (2018). Algorithms for interval-valued fuzzy soft sets in emergency decision making based on WDBA and CODAS with new information measure. *Computers & Industrial Engineering*, *119*, 439–452. DOI: 10.1016/j.cie.2018.04.001 PMID: 32288046

Qin, H., Ma, X., Herawan, T., & Zain, J. M. (2020). A novel soft set approach to decision making. *Knowledge-Based Systems*, *152*, 116–125. DOI: 10.1016/j.knosys.2018.04.018

Roy, A. R., & Maji, P. K. (2007). A fuzzy soft set theoretic approach to decision making problems. *Journal of Computational and Applied Mathematics*, *203*(2), 412–418. DOI: 10.1016/j.cam.2006.04.008

Santos, M. E., & Villatoro, P. (2018). A multidimensional poverty index for Latin America. *Review of Income and Wealth*, *64*(1), 52–82. DOI: 10.1111/roiw.12275

Weeks, J. R., Getis, A., Hill, A. G., Gadalla, M. S., & Rashed, T. (2020). The fertility transition in Egypt: Intraurban patterns in Cairo. *Annals of the Association of American Geographers*, *94*(1), 74–93. DOI: 10.1111/j.1467-8306.2004.09401005.x

Zadeh, L. A. (1965). Fuzzy sets. *Information and Control*, *8*(3), 338–353. DOI: 10.1016/S0019-9958(65)90241-X

Zhang, H., Xiong, L., & Ma, W. (2020). On interval-valued hesitant fuzzy soft sets. *Mathematical Problems in Engineering*, *2020*, 1649360. Advance online publication. DOI: 10.1155/2020/1649360

Chapter 6
Mapping Working Poverty from Space:
A Geospatial Approach to Socioeconomic Inequality

Ayesh Farooqi
http://orcid.org/0009-0000-4556-4211
Poornima University, Jaipur, India

Nikhil Kumar Goyal
http://orcid.org/0009-0007-4532-8033
Noida International University, Greater Noida, India

ABSTRACT

Applications The mapping of working poverty using geospatial tools has become a paradigm shift in terms of the understanding of the spatial distribution of the socioeconomic inequality in different regions. Combining satellite imagery, remote sensing indicators, GIS-based analytics, and social economic data, it is possible to reveal unobservable patterns of deprivation, labor concentration in the informal sector, and the differences in access to the most vital services. This chapter addresses the challenges in visualizing poverty in small spatial scales, highlight high-risk areas and analyzing the complicated intersection of environmental, demographic and economic variables which shape working poverty using geospatial

DOI: 10.4018/979-8-3373-7118-4.ch006

tools. The chapter shows how spatial intelligence can facilitate data-driven intervention strategies, strengthen social protection systems, and be used to promote inclusive development planning through case studies, analytical frameworks and discussions on the ethical implications of the spatial intelligence.

1. INTRODUCTION

1.1 Understanding Working Poverty in Contemporary Economies

Working poverty is one of the most undiscovered and long-lived problems in the world development. In comparison with absolute poverty, which is largely linked with unemployment or income deprivation, working poverty embodies more of a paradox despite the fact that the working poor are still employed and cannot meet basic living standards. This is because of poor remuneration, informal working conditions, unreliable working schedules and access to the social protection facilities. This has been worsened in much of the developing economies by the presence of informal or precarious work which has created a developing group of the working population languishing in hardships despite trying to actively contribute to the labor market. Working poverty requires understanding to create specific interventions that go beyond job-making that are further than merely securing jobs and mitigating structural inequities.

Winding up the complexity of being poor is the multi-dimensionality of poverty which is not only about earning money but also about being able to study, receive health care, live in a good house, and get the ability to move. Conventional measures of poverty do not fully measure these dimensions due to the significant dependence of surveys of households and self-reported data which can be out of date, incomplete or geographically restricted. As a result, policymakers can hardly determine the location of working poverty concentration and the community that is the most vulnerable. Spatial perspective thus turns out to be important in closing these data gaps in terms of interrelating socialeconomic conditions to physical landscapes, physical environment and settlement patterns (De la Barrera et al., 2016).

The geospatial analysis provides a paradigm change in understanding the working poverty as it allows the researcher to present economic vulnerability to a visual manner that cannot be achieved through a traditional dataset. Through mapping working poverty space and time, scholars are able to find the geographically persistent areas that are always underserved or disproportionately affected by economic shocks. Other spatial viewpoints that are also evident in this venture are the correlation between poverty and environmental hazards, transport system, urbanization, and markets and renewed accessibility. Consequently, geospatial intelligence provides the possibility to devise comprehensive, information-based policies that would target the underlying factors of working poverty and increase the resilience in the long run (Van Herzele & Wiedemann, 2003).

1.2 The Geographic Patterns of Deprivation and Geography

The concept of spatial inequality can be defined as imbalance in the allocation of economic opportunities, resources and conditions in different geographical areas. The poverty is not evenly distributed across all countries although in most of the countries it creates clustures and this happens to concentrate the poverty around environmental features, past settlements or unfair policymaking. These spatial inequalities can be mapped, which helps one gain a better insight into the role of structural factors in determining working poverty. As an example, the rural regions that are remote might not be connected with transport systems or formal employment, and urban unofficial regions can be overcrowded, poorly infrastructured, and exploited. The visualization of those trends using spatial data helps see how geography can either put people on the path to upward mobility or keep them in a state of hardship.

Figure 1. Geographic patterns of deprivation visualized through spatial inequality mapping

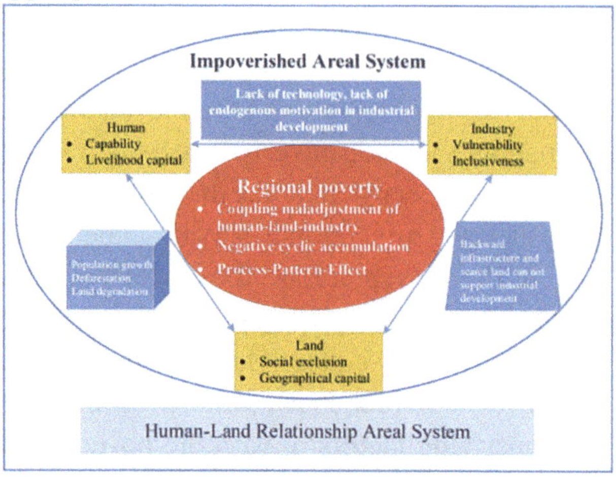

The geography of deprivation tends to be intersected when there are wider socioeconomic and environmental susceptibilities. Areas which are susceptible to natural shocks or weather changes like drought-prone farmlands or communities vulnerable to rising sea levels can have an increased amount of working poverty because of unpredictable means of earning and economic vulnerability. On the other hand, regions that have a well-developed infrastructure, in terms of good transport facilities, educational facilities, and business centers, are likely to have low rates of working poverty. To measure this relationship, geospatial tools can be used to correlate satellite derived indicators (e.g. vegetation health, land cover change or built-up density) with economic variables to discover the interaction between environmental and spatial variables and labor market (ISO, 2014).

The inequalities in government investment and priorities of policy may also be understood based on these geographic patterns. Areas characterized by chronic working poverty usually experience chronic abandonment, a dearth of access to governmental services and lack of social protection covers. With the help of spatial analysis, these disparities can be revealed and policymakers can address the resources towards the underserved zoning areas more accurately. Additionally, the location of clusters of poverty can be used in the process of local intervention measures like skills training, building infrastructures, and provision of subsidies. Governments will be able to develop more equitable policies which reflect specific circumstances of a particular area through incorporating geospatial insights into their formulation (Yao et al., 2014).

1.3 Relevance of Remote Sensing and GIS in the Socioeconomic Research

The analysis of socioeconomic inequality has been transformed by remote sensing and Geographic Information Systems (GIS) because of the ability to access data in high-resolution, timely, and scalable data. The satellite images help scientists to conclude on economy using the observable physical features in the surface like lights at night, use of land, quantification of buildings, and transport. These proxies can be of invaluable information on places where the traditional data could be limited, old-fashioned, or faulty. Indicatively, the intensity of light at night frequently is associated with economic activities in the area, and as such, it can be used to estimate the extent of incomes or identify the increase in informal settlements with a high degree of precision. These tools give a fresh perspective to study working poverty in small spatial scales which allows more detailed analysis and making decisions based on evidence.

GIS gives the computational framework of integrating indicators derived by satellites with traditional socioeconomic data. With GIS, combining census data, survey, environmental parameters, and spatial boundaries, it is possible to create a finer map about poverty that identifies the hotspots of working poverty. These applications enable a researcher to conduct some spatial statistics, identify the patterns of clustering, and examine the relationship between poverty and other geographic variables. GIS is also instrumental in scenario modeling to make policymakers know how future investments in infrastructure or policy changes in the future can affect various communities. This capability of modeling the outcomes improves future planning and encourages reasonable allocation of resources (Zhou & Kim, 2013).

Remote sensing and GIS are not merely useful in mapping, it enhances effectiveness and precision of tracking socio-economic developments. With the increasing frequency of satellite imagery becoming more and more frequent and high-resolution, scholars are able to monitor shifts in the levels of poverty throughout the years- enabling them to make decisions in real-time and react quicker to economic shocks. This is especially good in areas under conflict, in case of displacement, or natural calamities when it is hard to collect data on the ground. Eventually, remote sensing amid integration by GIS bolsters the world ability to study working poverty, allowing the development strategies to be more inclusive on congruence with the development sustainable agenda.

2. DATA SOURCES AND GEOSPATIAL INPUTS FOR POVERTY ANALYSIS

2.1 Satellite Imagery and Earth Observation Datasets

The foundation of geospatial analysis of poverty is provided by satellite images, which provide high-resolution information on surface phenomena related to the level of socioeconomic situation. Current planetary monitoring systems, like Landsat, Sentinel, MODIS, and commercial data services like Planet labs and Maxar, are able to capture fine spectral and spatial data to allow a researcher to track land use patterns, vegetation health, urban growth, and settlement structures. These aesthetic and spectral markers are strong proxies of the economical activity, availability of infrastructure, environmental stress, and all of them affect the dynamics of the working poverty. Continuous time-series satellite data can also be used to perform longitudinal evaluation, which can be used to see how economic trend changes, and how society reacts to events like natural tragedies, war, or policy change.

Remote sensing data is mostly helpful in the areas where social-economic surveys are not common or too old. Considering the example of low-income or politically unstable country, there are significant challenges posed to the established methods of poverty measurement by the impediments of logistics and the aspects of costs and security. The method of satellite systems overcomes these restrictions with the provision of standardised, repeatable and generally comparable datasets around the world. They also take into account marginal clues like roofing materials, road systems, farm subsidiary patterns or the rural occupancy which indicates affluence or the work conditions of the households. These features can be subjected to the quantitative measures of machine learning models that predict the levels of poverty using a greater level of accuracy (La Rosa, 2014).

Radar and thermal imaging are also used in addition to optical sensors to add special features to poverty mapping. Synthetic Aperture Radar (SAR) is not affected by the presence of cloud cover; moreover, it works during all types of weather, and its application is particularly effective in tropical areas when the optical images are frequently hard to see. Thermal sensors outline heat patterns that may indicate industrial activity, the quality of housing or the patterns of energy use. The aggregation of these multispectral and radar datasets offers a complete picture of the physical environments that give rise to working poverty and this allows multidimensional studies that transcend beyond the traditional socioeconomic data (Wright-Wendel et al., 2012.

2.2 Night-Time Light, Land use and environmental Indicators

Night-time light (NTL) imagery is one of the most common proxies in the measurement of the economic development and poverty. NTL data is derived mostly as a result of Defense Meteorological Satellite Program (DMSP) and Visible Infrared Imaging Radiometer Suite (VIIRS) and this data is a form of the artificial lighting produced by cities, industrial zones, and transport systems. Research has provided consistent results of high correlations between light intensity and other variables which include income, density, labor productivity and availability of infrastructure. Consequently, NTL datasets are often being used in the attempts to approximate the spatial distribution of working poverty, especially in areas where official statistics are low or unreliable. The temporal continuity further allows researchers to trace the changes in economic conditions because of alterations in policies, migration, or disturbance of the environment.

Another important geospatial input of poverty analysis is land use and land cover (LULC) datasets. These layers categorize regions into residential, commercial, agricultural, forest or barren land and provide an understanding of the way communities engage with the environment. The development of urban lands, slums, and agricultural stress can also expose socioeconomic vulnerability particularly where the populations are relying on informal or climatic-sensitive livelihoods. As an example, in most cases working poverty is concentrated in the regions that experience degraded agricultural land since agriculture lacks productivity and the access to the alternative areas of employment is limited. Likewise, the case of informal working markets and fragile lives has been exposed by unplanned urban settlements, in which urban patterns of spatial exclusion are manifested (Maas et al., 2009).

Remote sensor-based environmental indicators, including vegetation indices, drought severity, flood risk maps, and temperature anomalies also count in poverty analysis since they pointing out the ecological strains of low-income workers. The stressors in the environment have a direct impact on agricultural workforce, productivity of the informal sector, and household ability. Combined with socioeconomic, the indicators can be used to point to those regions where ecological vulnerability supports working poverty. The integration between environmental and economic data provides a more comprehensive perspective on climate change and environmental degradation together with the socioeconomic percentage variance (Gong et al., 2013).

2.3 Socioeconomic, Census, and Household Survey Data Integration

Although household-level data and socioeconomic data are critical in anchoring geospatial poverty measures to reality, satellite and environmental images are also practical in offering important proxies. The census documents, survey of labor, and data of home income, and demographic statistics present first hand information on the types of jobs, wages, education, and standard of living. These datasets serve as a supplement to the geospatial indicators because of the presence of contextual richness to understand the satellite-derived data appropriately. As an example, the differences in night-time lighting intensity can be the indicator of the economic activity, yet without income or employment statistics it is hard to discriminate between formal and informal labor markets or between high-output industrial areas and low-wages communities.

It is possible to integrate survey and satellite data to develop hybrid models which will greatly enhance the accuracy of poverty estimation. One can get machine learning algorithms matched with survey-based poverty labels and satellite-based features, since both generate copyright poverty maps at smaller spatial scales than traditional survey techniques are capable of. The strategy has worked in numerous applications in poverty mapping activities across the world and models using this method have the ability of pin pointing poverty hotspots with impressive accuracy even in data sparse areas. Such hybrid models are especially beneficial when examining working poverty as it usually varies over short geographic scales by interpolating between the known survey locations and remote sensing entities (Niemelä, 2014).

Additionally, using socioeconomic data in conjunction with geospatial inputs helps policy makers to come up with more specific interventions. As an illustration, through the combination of information about labor markets and the layers of environmental risks and access to transport, the policymakers can highlight the community where the working poverty involves a combination of various underlying factors. This type of multi-layered analysis helps the strategic investment in social protection, skills and development, infrastructure and movement of labour. Finally, survey and geospatial data integration would guarantee that poverty mapping is scientifically and socially relevant, which will provide a solid basis of policy-making and equitable developmental planning.

3. METHODOLOGICAL FRAMEWORK FOR GEOSPATIAL POVERTY MAPPING

3.1 Data Acquisition and Source Integration

The geospatial mapping approach to poverty analysis starts by ensuring that pertinent datasets are obtained in the socioeconomic data and the remote sensing platform. The spatial basis of this analysis is given by satellite imagery which can give multispectral, temporal and geolocated data capable of reflecting the features of the environment or the infrastructures over the areas where human activity takes place. Such frequently used platforms as Landsat, Sentinel-2, and VIIRS are chosen due to the best blend of resolution, availability, as well as spectral variety. These satellite data are used to identify proxies that are associated with poverty measurements such as built-up density, vegetation features, transportation system, nighttime lighting, and land-use change patterns. The remote sensing data cannot explicitly reflect the complexity of working poverty, thus; it has to be combined with the detailed socioeconomic data (Taylor et al., 2011)

The socioeconomic data are mostly acquired by accessing national census documents, labor force surveys, household income surveys and administrative records of employment. The variables found in these datasets include the household income, the type of employment, level of education, ratio of the dependency, and the demographic factors- data that is important in labelling and verifying the spatial poverty models. All the datasets are brought to the same temporal reference frame in order to be reliable. Congruency of time, format, the scale of space and separation of authority guarantees that the indicated points of remote sensing correspond to the actual social circumstances. After all, it is this integrated base of data that enables the geospatial models to embody the visible physical landscape and the invisible world of socioeconomic realities of the working families.

3.2 Data Preparation Data Preprocessing and Feature Extraction

After obtaining the necessary datasets, they get preprocessed to a large extent to achieve an analytical accuracy. The satellite imagery is also subject to radiometric correction, cloud masking as well as atmospheric adjustments that remove spatial distortion due to the environmental noise. Image sets are produced by uniform and region-specific using mosaic generation and spatial clipping. Upon preprocessing, measurements of a number of spectral indices and spatial metrics are calculated to identify contentful information of do-not-reside precipitation that is pertinent to working poverty estimation purposes. Vegetation indicators like the NDVI are useful in revealing agricultural productivity which is uber vital in areas where agricultural labor prevails. On the same note, developed indices, including NDBI, impervious surface measurements, provide an indication of the level of urbanization and industrialization, which, consequently, indicate access to formal jobs.

Night time light intensity as a derivative of VIIRS or DMSP-OLS can be a good proxy of economic events, electrification and availability of infrastructure. Spatial metrics and object-based image analysis methods are applied in extracting density of roads, distance to transportation networks and morphology of the settlement. The features of interest extracted are georelated and matched against socioeconomic data with shared spatial units like a grid cell, census tract, or a county or city. The given integration process will lead to a harmonized geospatial dataset in which each geospatial entity will be stored with both remote sensing indices and ground-truth poverty data. The ability of such detailed feature sets to cross-tabulate among the levels of various facets provides researchers with an opportunity to find out some hidden connections between working household life experiences and physical surroundings (Tian et al., 2014).

3.3 Spatial Modeling and Machine Learning Solutions

The last element of the methodological framework is the use of spatial modeling technique to forecast and map working poverty. Possible models of analysis that are likely to be applied when studying the relationship between spatial characteristics and poverty indicators are traditional statistical models Ordinary Least Squares (OLS) and Geographically Weighted Regression (GWR). OLS helps to obtain the global view of the impact of the factors in predicting the levels of poverty whereas GWR considers the local changes as it is clear that the causes of poverty are different in geographical settings. These models are interpretable and effective in learning the spatial heterogeneity, however, their effectiveness can be limited especially in complex and nonlinear environments.

In order to address these drawbacks, the geospatial poverty research has witnessed the growing popularity of machine learning and deep learning models. Random Forest, Support Vector Machines and Gradient Boosting algorithms are algorithms that can be particularly used to process large and high-dimensional data with nonlinear trends. They are able to combine various characteristics observed in satellite pictures to produce the accurate and trustworthy predictions of poverty. This area has further been advanced with deep learning algorithms, particularly the Convolutional Neural Networks (CNNs) that allow learning the pixel values of the raw image. The models of CNN based poverty estimation that utilize night time lights and multispectral images show high accuracy in geographic areas with scarce data and therefore assist the researchers predict poverty without conducting large field surveys (Schipperijn et al., 2010).

In a crucial aspect to take, spatial validation is the resulting step and becomes necessary to ensure that the generated poverty maps are robust and meaningful. The methods like spatial cross-validation, Moran I analysis, and residual mapping are employed to prove that the projected patterns are based on actual spatial groups of deprivation or a methodological artefact. Through testing outcomes in various regions and periods, researchers guarantee that the findings are applicable and reduce biasness. These modelling and validation processes reduce the raw geo spatial data to insights which can be acted on, understanding inequality and evaluating resource distribution as well as crafting specific intervention towards populations vulnerable to working poverty.

4. IDENTIFYING SPATIAL PATTERNS OF WORKING POVERTY

4.1 Detecting Geographic Hotspots of Economic Deprivation

To determine spatial patterns of working poverty, the first step is to identify geographic hotspots, or places where the rate of poverty is much greater than the one in neighboring areas. Hotspot analysis has been based on the geospatial statistical techniques to include: Getis-Ord Gi + index, Local Morans I, and kernel density estimation to establish areas of chronic deprivation. The tools assist in differentiating between the case of randomly spread poverty and one focused, which could be due to structural, environmental, or social factors. This kind of analysis indicates that working-poverty is not really an isolated phenomenon but rather is manifested in a discernible geographical localization, and is a resultant effect of demographic stressors, ineffective infrastructure, and a relative lack of economic diversification (Karteris et al., 2016).

Such hotspots areas may also be where peri-urban settlements, remote rural community or areas with a shrinking industrial base are found. The inability to match population expansion and job creation in the fast unfolding urbanising regions creates pools of working poor, who reside in high density informal settlements. On the other hand, hotspots can also manifest themselves in areas with soil erosion, weather strains or reliance on low-yield production in agricultural zones. These regions have been and, still, have low economic mobility, and hence the detection of spatial hotspots will prove invaluable in developing area-specific forms of development interventions. Visualisation of the spatial disparity on a granular level can enable the policymakers and practitioners to determine the areas that require resources the most.

4.2 Interpretation of Urban Rural Inequality in Working Poverty

An important aspect of spatial analysis is through the analysis of inequality between cities and the countryside. The access to services, education and employment opportunities are generally greater in urban areas though working poverty continues to increase as a result of increasing costs of living, informal labor and inequitable access to formal labor markets. Geospatial methods assist in disaggregating these complexities by map work of the distribution of different forms of employment, housing necessity, access to transportation and accessibility to services. These spatial characteristics when combined with poverty measurements show that the working poor in cities are frequently in segregated areas distant to economically oriented areas resulting in a longer commute and greater cost of living expense.

Figure 2. Visual representation of urban–rural disparities highlighting differences in living conditions and working poverty

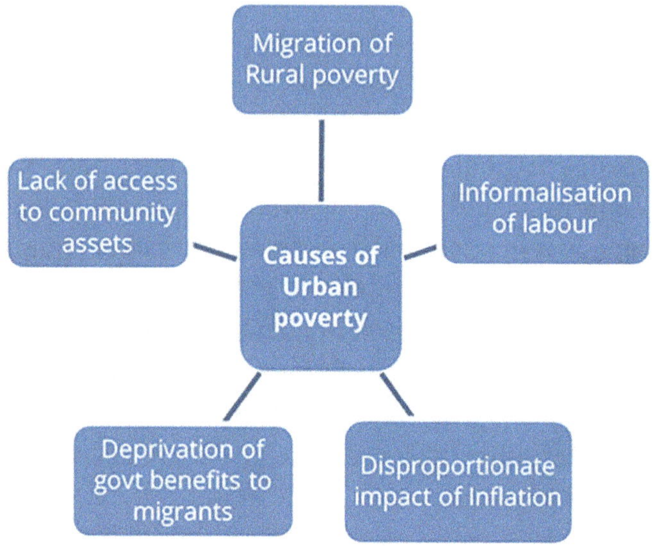

The causes of working poverty in rural areas are very different. Such reliance on agriculture subjects the workers to climate changes, variances in the market and not adopting new technologies. These underlying structural constraints are captured by satellite-based indicators (i.e. vegetation health, soil quality, and distance to roads). On space analysis, the rural poverty clusters tend to be related with agrarian stagnation, value-chain traps, and poor access to credit or agricultural extension services. The following insights enable the researcher to differentiate between structural and spatial determinants of poverty and can, therefore easily, suggest specific interventions, i.e., infrastructure development in the rural areas or urban livelihood support programs. The analysis of these two contexts through geospatial methods indicates that inequality is stratified and populations of working poor experience dissimilar levels of pressure within diverse contexts (Banzhaf & Kollai, 2015).

4.3 Experiencing Socioeconomic Inequality With Spatial Layers

Geospatial analysis is possessing power because it can help to visualize hidden social realities not seen in conventional tabular data. Multi-layer mapping allows the researchers to map the poverty levels in combination with the spatial variables, education levels, access to health facilities, and ease of transportation and the quality of homes. These layers aid in exposing systemic problems in terms of inequalities and spatial imbalance of resource allocation. An example is plotting poverty and education infrastructure that can indicate that a working-poor household in an area with limited number of schools or a lower number of teachers. Likewise, map and superimposition of poverty and health facility location can give information of the places where workers can be more vulnerable since they lack access to basic services (Hou & Walz, 2014).

Such stratified visualizations also reveal inequality in investment on the infrastructure. More affluent neighborhoods in most places have dense road systems, regular electricity, and brighter lights at night, while working-poor ones are characterised by disrupted transport networks, low levels of illumination. These visual contrasts are meaningful in displaying uneven distribution of flow of development and the services to the people. Maps of spatial inequality also provide the association between environmental risks and poverty. These communities of working poor tend to be sited in such areas where floods are likely to occur or where there might be a drought or land degraded conditions- all this limits economic development. Mapping these overlapping vulnerabilities will enable the policymakers to foresee risks that will be merged together and develop the solutions that will be spatially informed. Multi-layer mapping makes the multifaceted character of the socioeconomic inequality more clear and allows allocating resources in an evidence-based manner (Banzhaf & Höfer, 2008).

4.4 Time-Varying Mapping and Change in the Long Run

Besides the spatial patterns, it is essential to understand how working poverty evolves with time so as to produce sustainable solutions. Temporal mapping will use satellite records of time series to identify land-use change to urban growth, agricultural productivity, and development of nighttime lights. Utilizing the geospatial poverty maps of several years, the researchers are capable of establishing whether some areas are in a slow transition up or down or those areas are in persistent deprivation. As an example, the brightness of nighttime lighting can be used to associate with the possibility of an increase in economic activity and the growth of informal settlements can be used to suggest deteriorating conditions of the working poor. This time lens can enable analysts to draw the line between minor changes with a short-term nature and structural changes that take place over a long period.

Longitudinal analyses also assist in testing the policy interventions. Spatial changes in the clusters of poverty can be quantified using temporal geospatial models when new infrastructure development, social welfare programs and economic development projects are put into place. Although with time poverty hotspots may decrease, that would be a positive indication of successful policies; but on the other hand, increasing hotspots can be a reminder that there is a gap in policy, or that the policy will have counterproductive effects. On the same note, vegetation variation over time aids in monitoring agricultural areas where climatic variation poses a threat to human livelihood thus allowing areas likely to be subjected to working poverty in the future to be identified early enough. Comprehensively, temporal mapping provides a dynamic level of analysis of the geospatial poverty, enabling the researcher to develop a way of knowing not only where poverty is; but also how and why it changes with time (Banzhaf et al., 2013).

5. LINKING GEOSPATIAL PATTERNS TO POLICY AND GOVERNANCE

5.1 Using Spatial Intelligence to Identify Priority Intervention Zones

Geospatial analysis is key in providing priority areas where the policy intervention could have most effects in alleviating working poverty. The conventional measures of poverty do not have the spatial fineness to identify small-scale inequalities on the district or municipality level. Conversely, high resolution maps of satellite images and mapping of the same through GIS, sheds light of dependent groups of deprivation, and as such, this allows the policymakers to see where economic vulnerability gathers in the greatest density. Such lessons enable governments to abandon the wide approaches to interventions that are narrow and homogenous and embrace those tailored in-depth approaches that allocate resources more effectively and equitably.

In the case, maps that depict a lack of transportation infrastructure access will inform investments in roads, or mass transit or micro-mobility solutions that will subsequently increase the employment connectivity among working-poor population groups. In the same fashion, the spatial models which indicate poor accessibility to clean water, electricity or sanitation facilities can be used to design infrastructure development plans that are unique to the disadvantaged regions. Governments and development organizations will be able to transition to evidence-based planning by incorporating spatial intelligence into policy development and reducing any form of inefficiency by preventing mishaps and maximizing social impacts (Romero et al., 2012).

The identification of priority zones using geospatial is also useful in anticipating the risk in the future. Regions that are vulnerable to climatic risks, including agricultural belts that are susceptible to drought or urban settlements that are susceptible to floods can be marked as potential preventive actions. These interventions can be in the form of climate resistant agricultural measures, disaster preparedness measures or adaptation housing measures. By so doing, spatial intelligence not only recognize the current patterns of poverty, but also anticipates where the other areas of vulnerability are likely to emerge such that the policy interventions are proactive and sustainable (De la Barrera et al., 2016).

5.2 Ameliorating Social Protection Systems using Geospatial Decision Support

Geospatial decision-support tools are very useful in modern social protection systems to assist governments in the identification of beneficiaries, resources allocation and monitoring program implementation. The conventional welfare targeting approaches tend to depend on old surveys, self-reported information or general administrative lines that do not reflect the real picture of working poverty. Geospatial poverty mapping adds to these systems with near-real-time or real time data available on economic vulnerability, housing conditions, and disparities, which includes infrastructures. This integration will enable welfare programs to provide the most effective support of the households in need.

A significant benefit is the integration of geospatial databases with administrative databases that may be tax registries or social welfare databases. Once the socioeconomic profiles are combined with spatial properties, such as risk factors that are typically location-based or neighborhood development indicators, policymakers will have a better view of household vulnerability. Such detail is a treasure trove of conditional cash transfer programs, food security projects, or employment guarantee programs that need to be correctly targeted to avoid errors of exclusion and minimize leakage.

Besides, geospatial services have the ability to track the area covered and the efficacy of the social protection initiatives in the long run. Geographic mapping of the beneficiaries will help the policymakers to see the areas that were underserved or discrepancies in the program execution. Environmental indicators can also be integrated in the spatial dashboards and noted of the areas where climate shocks in combination with poverty require rapid response settings. Since social protection is becoming more and more data-driven, geospatial intelligence is an essential solution to a reliable and inclusive and adaptable welfare system (Albehery & Wang, 2011).

5.3 Developing Planning and Infrastructure Investment by improving Spatial Planning Analysis

Planning of development and investment in infrastructure has been overtaken by spatial analysis in order to know where economic development activities will give the most returns. The presence of geospatial poverty maps can reveal discrepancies between population requirements and the current structure so that it is possible to plan schools, healthcare facilities, energy networks, transportation networks, and digital connectivity infrastructure in a more efficient way. Combined with the indicators of poverty, these maps indicate areas where the working-poor families do not have access to basic services, which would enhance their socioeconomic mobility.

The spatial analysis also helps in the economic development strategies of the areas as they can indicate the areas where there is a potential that has not been fully exploited. As an example, the areas with middle level of poverty and high agricultural productivity would be able to enjoy an investment in processing plants or market connections. On the other hand, the regions containing low natural resources and great concentration of poverty might have to carry out alternative livelihoods programs, vocational centres or microenterprise development initiatives. The development planning process has to be based on the empirical spatial evidence, with the help of which policymakers may equip interventions based on the particular needs and opportunities of the specific region (Albehery & Wang, 2010).

Moreover, insights gained via geospatial opportunities enhance open and responsible decision making. Spatial dashboards, which are open and face the people, enable the civil society and the citizens to track how the infrastructure investments have met their requirements. They also aid in finding areas that had been overlooked in the past by dictation by political leaders or incompetent acts by administrations. Governments can make spatial information more democratic to lessen disparity in the distribution of public service to the population and achieve more inclusive growth. After all, geospatial analysis should be brought to development planning to make sure that investments not only contribute to economic growth but also inequality reduction and promotion of social justice (Albehery & Wang, 2011).

5.4 Promoting Multi- layers Governance and Co-ordination Across Sectors

The influence of working poverty depends on various factors, which are labor markets, environmental conditions, quality of education, and availability of social services. Good governance, thus, involves the harmonization of administration at various administrative tiers and regions. Geospatial analysis offers a standard information relationship that allows ministries, local governments, nongovernmental organizations, and international agencies to cooperate and coordinate their intervention. The spatial datasets can assist in addressing the institutional gap, providing objective evidence of a location, which the entire stakeholder group can interpret and apply to make decisions.

An example is that national governments can use geospatial maps of poverty to distribute funds to states or cities, and local governments can use smaller-scale maps that are based on neighbourhoods to plan community-based initiatives. Geospatial knowledge on the other hand can help development agencies in coordinating any type of humanitarian aid, climate resilience, or livelihood support in areas at risk. By common geospatial platforms, stakeholders would keep track of progress that would not be repeated and would increase integration of the institutions (Zadeh, 1965).

Adaptive governance is also brought about by geospatial tools. The more current the data is released, be it updated satellite imaging, questionnaires conducted within households, or even real-time sensors, the more governments are able to adjust their strategies and resource allocation, and also work dynamically. This process enhances the governance frameworks and is performed on a cyclical basis such that policies are always responsive to the dynamic socio economical environments. Geospatial intelligence as it were, serves as a binding concept which facilitates multisectorial cooperation, institutional candor and ultimately leads to improved and fairer governance (Klir & Yuan, 1995).

6. CHALLENGES, LIMITATIONS, AND ETHICAL DIMENSIONS

6.1 Data Quality and Resolution Constraints in Geospatial Poverty Mapping

Among the most important issues regarding the use of the geospatial data in mapping the working poverty is the difference in the data quality, spatial resolution between different regions and data sets. An example of satellite imagery is between rough-resolution global output and fine-resolution commercial output. Although high-resolution features may be used to provide high-detailed data as the extent of building density or quality of infrastructure, it is frequently costly and not always applicable across the geographic location. Reduced resolution images, however, might fail to capture essential variations at the micro-level, which determine patterns of poverty, notable where poverty is highly concentrated in large urban capitals where the spatial ranges of socioeconomic statuses relative to each other can be phenomenally different.

Also, the unequal data collection procedures in different countries and institutions lead to the unequal results that make it difficult to compare the results across regions. The remote sensing variables like the lights at night are potentially useful in urban environments but not so dependable in the countryside where economic activity is not necessarily linked to man-made lights. Equally, land-use data can be inaccurately defined in multifaceted topography, and thus the causes of working poverty are not accurately interpreted. These issues make the significance of stricter data validation, harmonization developments, and ground-truthing procedures apparent because this way, geospatial models of poverty can be used to incorporate the actual situation on the ground.

Geospatial mapping is also complicated by shortcomings in time. Time-series no longer observations due to satellite revisiting, cloud cover, seasonal factors and incomplete update intervals may lead to gaps in satellite others datasets. These gaps are barriers to monitoring changes in the state of poverty or any kind of early warning of new vulnerability. In the absence of regular and consistent time-based data, longitudinal analyses can give false trends or fail to consider any short sense shock that has a disproportionate effect on working-poor populations. The way to overcome such constraints is to invest in data infrastructure, open-access satellites, and built-in data pipelines that will facilitate continuous and quality-monitoring (Nguyen & Walker, 2006).

6.2 Algorithmic Bias, Interpretation Problems and Representation Risk

Machine learning and AI in geospatial analysis of poverty also present new threats of bias in the algorithm and data misinterpretation. Training models based on biased data may solve the discrepancies unknowingly by linking some geographical features with poverty. As an illustration, the models might overstress visibly distinguishable items like the building size or the rate of the road density and these items may not vary along the economic lines or even in line with some other construction or building cultures. When not calibrated, these correlations may result in a wrong assessment of poverty or misidentification of a community, which may have implications on the distribution of resources.

The issues of interpretation can occur because policymakers base their choices on the geospatial outputs without having a deeper comprehension of the assumptions and limitations that they carry. Although poverty maps appear to be more visual, they reduce the intricate socioeconomic reality into formal patterns of space that can contain structural problems of the past (i.e., historical discrimination), gaps in governance, or labor market failures. This excessive dependency on far-off sensing proxies can also marginalize the relevance of the qualitative value of the community involvement or an ethnographic study. Consequently, policymakers need to integrate geospatial information with terrestrial, situational information in order to prevent misinterpretation and to make interventions culturally-sensitive and place-based (Nguyen & Kreinovich, 1997).

The problem of representation also is very important in geospatial ethics. The categorization of an area as a poverty hot spot will have the adverse effect of stigmatizing the communities it targets without their intention and could lead to social or political backlash. These labels should be effectively expressed and there should be focus on structural causes as opposed to community attributes. Also, the fact that high-resolution imagery is used brings up the issue of visual surveillance and the possibility of revealing sensitive data of the marginalized. The ethical geospatial activities hence involve transparency, informed consent where necessary and protocols that safeguard the expectations and entitlement of vulnerable communities.

6.3 Data Privacy, Confidentiality and Consent in Geospatial Socioeconomic Mapping

This is of particular importance when geospatial analysis can comprise such a mixture of satellite and micro-level socioeconomic data as household surveys, mobile data, or administrative records. Even though satellite data does not necessarily include any personal indicators, many layers combined can allow one to conclude about some sensitive information about a person or a small group. The question of this risk is how governance of data, compliance with the law, and the ethics of spatial poverty studies. The policymakers have to make sure that the data collection and integration processes adhere to powerful privacy frameworks so that they are not exposed or manipulated unintentionally.

There is yet another level of complexity that is brought about by the unequal nature of regulation among various countries. Laws in data protection differ in many aspects, which gives the inconsistent standards of consent, data sharing, and anonymization. It is delicate to overcome these differences by researchers lest they contravene local laws or global privacy standards. Some of the tools that may be utilized to minimise the risk of re-identification are spatial masking, aggregation, and differential privacy that also lead to compromised spatial precision and analysis quality. The problem of privacy and analytical accuracy is an issue that continues to plague geospatial poverty mapping (Conservation Measures Partnership, 2013).

In geospatial studies of large scale, consent is usually neglected. Whereas satellite and remote sensing data do not necessitate face-to-face interaction with individuals, socioeconomic datasets utilized with them will, in most instances, involve human subjects. When the data is reused to achieve other purposes than it was initially intended then the guarantee of meaningful consent is hard to achieve. The ethical procedures of research, therefore, entail clear communication concerning the manner in which the information will be applied, individuals who will view it, and the risks that the study-subjects will experience. Through the introduction of privacy protection, consent mechanisms, and efficient data governance into geospatial poverty studies, researchers and organisations can, nevertheless, fulfill ethical principles, yet still gain the full transformative power of spatial intelligence.

6.4 Limitations of Satellite-Based Socioeconomic Proxies

Even though satellite indicators (lights at night, vegetation index, and urbanized area estimates) are valuable sources of information, they are all a proxy and not a real measure of poverty. These proxies do not always provide a complete understanding of the complexity of working poverty which is influenced by the intangible factors, i.e. labour conditions, wage structure, social mobility and access to opportunity. A case in point is that the illumination of a place may rise as a result of electrification or commercial growth in the region yet the wages remain in stagnation or they may be exploited. The use of proxies therefore alone may result in an incomplete, or false image of the conditions of poverty.

Furthermore, satellite proxies may act variously in geographical and economic situations. An agricultural area with minimum light emissions can be economically stable, whereas an urban slum where light is high can represent high but low-income living conditions. Such discrepancies demonstrate the necessity to integrate both satellite information and quality socioeconomic data and perform the field adjustments to correct the interpretations. Another reason to have methodological transparency is proxy limitation, which should allow users to determine what geospatial datasets can measure correctly and the location of their interpretive limits.

CONCLUSION

Use of geospatial technologies to define working poverty has revolutionally reshaped the manner in which researchers, governments, and development organizations have comprehended social economic inequality. With satellite images, spatial analytics, and machine-enhanced mapping powered by machine learning it is now possible to visualize the ascertainment, seriousness and neglect behind working poverty more precisely than ever before. These instruments are able to light up space patterns which still were concealed in the conventional survey-based systems, and the space patterns are able to define the impacts of the environmental constraints, infrastructural inadequacy, and unequal economic development on instinctive realities of working-poor people. Urban-rural comparative analysis, space hotspots, and multi-layers mapping enable geospatial techniques to present the undivided view of deprivation both in its physical and socioeconomic manifestations.

Simultaneously, geospatial intelligence is invaluable when it comes to crafting more targeted, efficient, and equitable policies of the people. Spatial informatics can be applied to inform decisions at all levels of governance, including optimizing housing investments and social protection systems through pinpointing areas of priority intervention, as well as the more generalization of priorities. Through the connection of spatial evidence and development planning, governments are able to more effectively distribute their resources, respond faster to the arising vulnerabilities, and monitor all the long-term effects of their programs. This place-based strategy does not only make the policy interventions more accurately operationalized but also more inclusive to ensure that support is extended to communities that have historically been neglected or had been marginalized.

In spite of its potential to transform, geospatial poverty mapping has its own challenges, such as lack of quality data, bias in the algorithms, privacy issues, and reduction of proxies of satellites used in socioeconomic measurements. The means to face those pitfalls are a sound system of data governance, transdisciplinary work, and ethical conduct in research by honoring the dignity of the community and preserving vulnerable groups. With the emerging technology of geospatial technologies, their desirable and equitable utilization will be necessary in order to realize the potential of the technologies. Finally, space mapping working poverty can deliver more than an image, but this is also an opportunity to transform the strategies of development, to strengthen social justice, and to create pathways towards inclusive and sustainable prosperity.

REFERENCES

Albehery, N., & Wang, T. (2010). Statistical inference of poverty measures. *UTEP/NMSU Workshop*.

Albehery, N., & Wang, T. (2011). Statistical inference of poverty measures using U-statistics approach. *International Journal of Intelligent Technologies and Applied Statistics*, *4*(2), 245–263.

Banzhaf, E., & Höfer, R. (2008). Monitoring urban structures using CIR aerial photographs. *JSTARS*, *1*(2), 129–138.

. Banzhaf, E., Kollai, H. (2015). Monitoring urban tree cover with remote sensing. *ISPRS Archives*, XL-7/W3.

Banzhaf, E., Reyes-Paecke, S., Müller, A., & Kindler, A. (2013). Demographic & land-use change dynamics. *Habitat International*, *39*, 179–191. DOI: 10.1016/j.habitatint.2012.11.005

. Conservation Measures Partnership. (2013). Open Standards for Conservation Practice v3.0.

De la Barrera, F., Reyes-Paecke, S., & Banzhaf, E. (2016). Indicators for green spaces in contrasting urban settings. *Ecological Indicators*, *62*, 212–219. DOI: 10.1016/j.ecolind.2015.10.027

De la Barrera, F., Reyes-Paecke, S., Harris, J., Bascuñán, D., & Farías, J. M. (2016). People's perception influences on the use of green spaces in socio-economically differentiated neighborhoods. *Urban Forestry & Urban Greening*, *20*, 254–264. DOI: 10.1016/j.ufug.2016.09.007

Gong, C., Chen, J., & Yu, S. (2013). Biotic homogenization of flora in urban green spaces. *Landscape and Urban Planning*, *120*, 158–169. DOI: 10.1016/j.landurbplan.2013.08.006

Hou, W., & Walz, U. (2014). Extraction of small biotopes using multi-temporal remote sensing. *International Journal of Remote Sensing*, *35*, 7245–7262. DOI: 10.1080/01431161.2014.967890

ISO. (2014). ISO 37120: Sustainable development of communities – Indicators for city services and quality of life.

Karteris, M., Theodoridou, I., & Mallinis, G. (2016). Green roofs assessment using GIS and remote sensing. *Renewable & Sustainable Energy Reviews*, *58*, 510–525. DOI: 10.1016/j.rser.2015.11.098

Klir, G. J., & Yuan, B. (1995). *Fuzzy Sets and Fuzzy Logic*. Prentice Hall.

La Rosa, D. (2014). Accessibility to greenspaces: GIS-based indicators for sustainable planning. *Ecological Indicators*, *42*, 122–134. DOI: 10.1016/j.ecolind.2013.11.011

Maas, J., van Dillen, S. M. E., Verheij, R. A., & Groenewegen, P. P. (2009). Social contacts as a mechanism between green space and health. *Health & Place*, *15*(2), 586–595. DOI: 10.1016/j.healthplace.2008.09.006 PMID: 19022699

Nguyen, H. T., & Kreinovich, V. (1997). *Applications of Continuous Mathematics to Computer Science*. Kluwer. DOI: 10.1007/978-94-017-0743-5

Nguyen, H. T., & Walker, E. A. (2006). *First Course in Fuzzy Logic*. CRC Press.

Niemelä, J. (2014). Ecology of urban green spaces: Key research questions. *Landscape and Urban Planning*, *125*, 298–303. DOI: 10.1016/j.landurbplan.2013.07.014

Romero, H., Vásquez, A., Fuentes, C., Salgado, M., Schmidt, A., & Banzhaf, E. (2012). Assessing urban environmental segregation (UES). The case of Santiago de Chile. *Ecological Indicators*, *23*, 76–87. DOI: 10.1016/j.ecolind.2012.03.012

Schipperijn, J., Stigsdotter, U. K., Randrup, T. B., & Troelsen, J. (2010). Influences on the use of green space. *Urban Forestry & Urban Greening*, *9*(1), 25–32. DOI: 10.1016/j.ufug.2009.09.002

Taylor, B. T., Fernando, P., Bauman, A. E., Williamson, A., Craig, J. C., & Redman, S. (2011). Measuring the quality of public open space using Google Earth. *American Journal of Preventive Medicine*, *40*(2), 105–112. DOI: 10.1016/j.amepre.2010.10.024 PMID: 21238857

Tian, Y., Jim, C. Y., & Wang, H. (2014). Assessing ecological quality of urban green spaces. *Landscape and Urban Planning*, *121*, 97–108. DOI: 10.1016/j.landurbplan.2013.10.001

Van Herzele, A., & Wiedemann, T. (2003). A monitoring tool for the provision of accessible and attractive urban green spaces. *Landscape and Urban Planning*, *63*(2), 109–126. DOI: 10.1016/S0169-2046(02)00192-5

Wright-Wendel, H., Zarger, R. K., & Mihelcic, J. R. (2012). Accessibility and usability: Green space perceptions in Latin America. *Landscape and Urban Planning*, *107*, 272–282. DOI: 10.1016/j.landurbplan.2012.06.003

Yao, L., Liu, J., Wang, R., Yin, K., & Han, B. (2014). Effective green equivalent—A measure of public green spaces for cities. *Ecological Indicators*, *47*, 123–127. DOI: 10.1016/j.ecolind.2014.07.009

Zadeh, L. A. (1965). Fuzzy sets. *Information and Control*, *8*(3), 338–353. DOI: 10.1016/S0019-9958(65)90241-X

Zhou, X., & Kim, J. (2013). Social disparities in tree canopy and park accessibility. *Urban Forestry & Urban Greening*, *12*(1), 88–97. DOI: 10.1016/j.ufug.2012.11.004

Conclusion

The exploration of space-based technologies as instruments for poverty alleviation reveals both a profound opportunity and a critical responsibility. Throughout this volume, it has become increasingly evident that the condition of the working poor cannot be fully understood – or effectively addressed – through conventional approaches alone. Poverty, particularly among those who are employed yet economically insecure, is multidimensional, dynamic, and deeply embedded within spatial, social, and economic systems. By extending our perspective beyond traditional ground-based methods, space-based technologies offer a powerful means to illuminate these complexities with greater clarity, scale, and precision.

Satellite imagery, remote sensing, and global navigation systems have demonstrated their capacity to transform how poverty is identified and analyzed. By enabling the integration of geospatial data with socioeconomic indicators, these technologies provide a more nuanced understanding of where poverty exists, how it evolves, and which populations are most affected. The identification of poverty hotspots, infrastructure gaps, environmental vulnerabilities, and service disparities allows for more targeted and efficient interventions. In this way, space-based technologies contribute not only to improved data accuracy but also to more informed and strategic decision-making.

At the same time, this book has emphasized that technology, in and of itself, is not a solution. The true value of space-based innovations lies in how they are applied – how data is translated into action, how insights are aligned with local realities, and how interventions are designed to be inclusive, equitable, and sustainable. The frameworks presented in this volume – the Spatial Poverty Intelligence Framework (SPIF) and the

Inclusive Space Technology Application Model (ISTAM) - highlight the importance of bridging technical capability with human-centered implementation. Together, they provide a roadmap for leveraging space-based tools in ways that are both analytically rigorous and socially responsive.

The interdisciplinary nature of this work underscores the complexity of poverty as a global challenge. Contributions spanning geospatial mapping, image processing, domain adaptation, ethics, financial literacy, renewable energy, and medical imaging illustrate that meaningful solutions require collaboration across fields. Addressing the needs of the working poor demands not only technological innovation but also insights from social sciences, public policy, and community engagement. It is through this convergence of knowledge that more holistic and impactful strategies can emerge.

Equally important are the ethical considerations that accompany the use of advanced technologies. As access to satellite data expands and analytical tools become more sophisticated, questions of data privacy, representation, and equity must remain at the forefront. Ensuring that the benefits of space-based technologies are distributed fairly – and that vulnerable populations are not further marginalized – requires intentional governance, transparency, and accountability. The commitment to ethical practice is not peripheral but central to the legitimacy and effectiveness of these approaches.

Looking ahead, the role of collaboration will be pivotal. Partnerships among governments, academic institutions, private sector innovators, and international organizations will shape the future of space-based applications in development. Open data initiatives and capacity – building efforts will be essential in democratizing access to these tools, enabling a broader range of stakeholders to participate in data-driven decision-making. Empowering local communities with the knowledge and resources to utilize space-based technologies will be a key factor in ensuring long-term sustainability and impact.

This volume also points to the importance of adaptability. As global conditions continue to shift – driven by climate change, urbanization, technological advancement, and economic volatility – the challenges faced by the working poor will likewise evolve. Space-based technologies, with their ability to provide real-time and longitudinal insights, are uniquely positioned to support adaptive and responsive strategies. However, this potential can only be realized through continuous innovation, critical reflection, and a willingness to integrate new knowledge and perspectives.

Ultimately, *Space-Based Technologies in Poverty Alleviation for the Working Poor* calls for a reimagining of how we approach poverty in an increasingly interconnected and data-rich world. It challenges us to move beyond fragmented and reactive solutions toward more integrated, proactive, and evidence-based approaches. By harnessing the power of space-based technologies – while remaining grounded in the realities and needs of the communities we seek to serve – we can take meaningful steps toward reducing inequality and expanding opportunity.

The journey toward poverty alleviation is complex and ongoing, but it is not without hope. With the right combination of innovation, collaboration, and ethical commitment, space-based technologies can play a transformative role in shaping a more inclusive and equitable future. It is our hope that this book serves not as an endpoint, but as a catalyst for continued exploration, dialogue, and action in the pursuit of improving the lives and livelihoods of the working poor worldwide.

Figure 1. A chart to represent key impact areas of space-based technologies in poverty alleviation

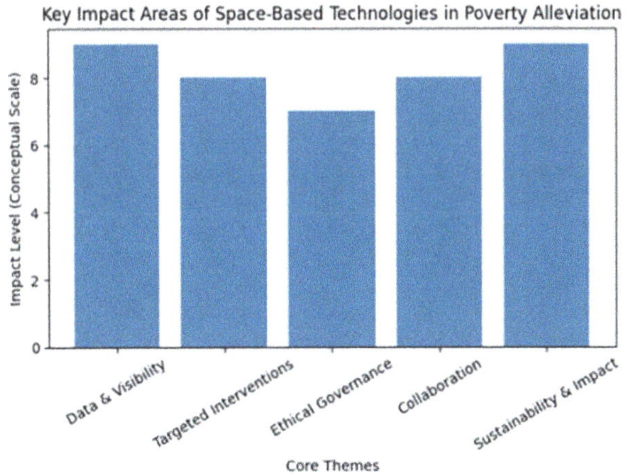

Figure 2. A systems diagram showing relationships between themes

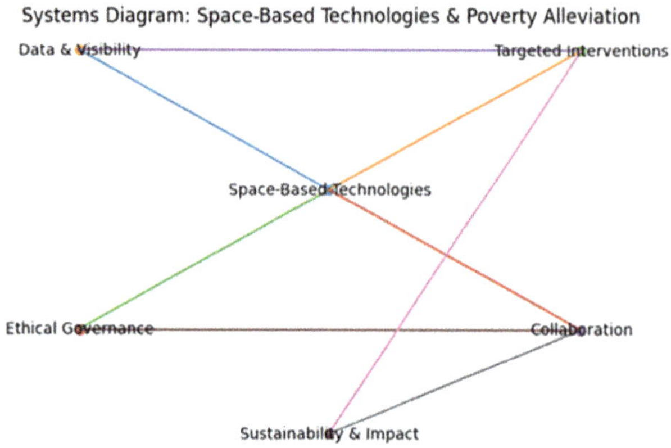

Compilation of References

Abbasi, M., Majdzadeh, R., Zali, A., Karimi, A., & Akrami, F. (2018). The evolution of public health ethics frameworks: Systematic review of moral values and norms in public health policy. *Medicine, Health Care, and Philosophy*, *21*(3), 387–402. DOI: 10.1007/s11019-017-9813-y PMID: 29124449

Aburas, M. M., Ho, Y. M., Ramli, M. F., & Ash'aari, Z. H. (2019). Improving the capability of an integrated CA-Markov model to simulate spatio-temporal urban growth trends using an Analytical Hierarchy Process and Frequency Ratio. *International Journal of Applied Earth Observation and Geoinformation*, *59*, 65–78. DOI: 10.1016/j.jag.2017.03.006

Aggarwal, P., Vyas, S., Thornton, P., Campbell, B. M., & Kropff, M. (2019). Importance of considering technology growth in impact assessments of climate change on agriculture. *Global Food Security*, *23*, 41–48. DOI: 10.1016/j.gfs.2019.04.002

Ahmad, B., & Kharal, A. (2009). On fuzzy soft sets. *Advances in Fuzzy Systems*, *2009*(1), 586507. Advance online publication. DOI: 10.1155/2009/586507

Akbari, M., Neamatollahi, E., Neamatollahi, P., & Akbari, E. (2020). Prioritizing municipal solid waste landfill site selection using fuzzy-AHP analysis. *Environmental Science and Pollution Research International*, *27*(21), 26283–26294. DOI: 10.1007/s11356-020-08929-y

Albehery, N., & Wang, T. (2010). Statistical inference of poverty measures. *UTEP/NMSU Workshop*.

Albehery, N., & Wang, T. (2011). Statistical inference of poverty measures using U-statistics approach. *International Journal of Intelligent Technologies and Applied Statistics*, *4*(2), 245–263.

Ali, G., Alolaiyan, H., Pamučar, D., Asif, M., & Lateef, N. (2021). A novel MADM framework under q-rung orthopair fuzzy bipolar soft sets. *International Journal of Fuzzy Systems*, *23*(7), 2163–2182. DOI: 10.1007/s40815-021-01095-7

Alkire, S., & Foster, J. (2011). Counting and multidimensional poverty measurement. *Journal of Public Economics*, *95*(7–8), 476–487. DOI: 10.1016/j.jpubeco.2010.11.006

Alkire, S., & Santos, M. E. (2013). A multidimensional approach: Poverty measurement & beyond. *Social Indicators Research*, *112*(2), 239–257. DOI: 10.1007/s11205-013-0257-3

Anselin, L. (2019). A local indicator of multivariate spatial association: Extending Geary's c. *Geographical Analysis*, *51*(2), 133–150. DOI: 10.1111/gean.12164

Atkinson, A. B., & Marlier, E. (2010). *Income and living conditions in Europe*. Publications Office of the European Union., DOI: 10.2785/53320

Avgerou, C. (2010). Discourses on ICT and development. *Information Technologies and International Development*, *6*(3), 1–23. https://eprints.lse.ac.uk/35564/

Banerjee, A., & Duflo, E. (2013). Poor economics. A radical rethinking of the way to fight global poverty'. *Society and Economy*, *35*(4), 573–587. https://www.ceeol.com/search/article-detail?id=622820

Banzhaf, E., & Höfer, R. (2008). Monitoring urban structures using CIR aerial photographs. *JSTARS*, *1*(2), 129–138.

Banzhaf, E., Reyes-Paecke, S., Müller, A., & Kindler, A. (2013). Demographic & land-use change dynamics. *Habitat International*, *39*, 179–191. DOI: 10.1016/j.habitatint.2012.11.005

Barentine, J. C., Walczak, K., Gyuk, G., Tarr, C., & Longcore, T. (2021). A case for a new satellite mission for remote sensing of night lights. *Remote Sensing (Basel), 13*(12), 2294. DOI: 10.3390/rs13122294

Barnett, C. (2018). Geography and the priority of injustice. *Annals of the Association of American Geographers, 108*(2), 317–326.

Barrington-Leigh, C. J., & Millard-Ball, A. (2017). The world's user-generated road map is more than 80% complete. *PLoS One, 12*(8), e0180698. DOI: 10.1371/journal.pone.0180698 PMID: 28797037

Baud, I., Sridharan, N., & Pfeffer, K. (2019). Mapping urban poverty for local governance in an Indian mega-city: The case of Delhi. *Urban Studies (Edinburgh, Scotland), 56*(5), 934–951. DOI: 10.1177/0042098007087341

Becker-Reshef, I., Justice, C., Sullivan, M., Vermote, E., Tucker, C., Anyamba, A., Small, J., Pak, E., Masuoka, E., Schmaltz, J., Hansen, M., Pittman, K., Birkett, C., Williams, D., Reynolds, C., & Doorn, B. (2010). Monitoring global croplands with coarse resolution earth observations: The global agriculture monitoring (GLAM) project. *Remote Sensing (Basel), 2*(6), 1589–1609. DOI: 10.3390/rs2061589

Benitez, K., & Malin, B. (2010). Evaluating re-identification risks with respect to the HIPAA privacy rule. *Journal of the American Medical Informatics Association : JAMIA, 17*(2), 169–177. DOI: 10.1136/jamia.2009.000026 PMID: 20190059

Betti, G., Gagliardi, F., Lemmi, A., & Verma, V. (2018). Comparative measures of multidimensional deprivation in the European Union. *Empirical Economics, 55*(3), 1071–1100. DOI: 10.1007/s00181-014-0904-9

Betti, G., & Verma, V. (2019). Fuzzy measures of longitudinal poverty in a comparative perspective. *Social Indicators Research, 142*(1), 249–272. DOI: 10.1007/s11205-018-1911-5

Brady, D., Blome, A., & Kleider, H. (2021). How politics and institutions shape poverty and inequality. In Brady, D., & Burton, L. M. (Eds.), *The Oxford handbook of the social science of poverty* (pp. 555–578). Oxford University Press., DOI: 10.1093/oxfordhb/9780199914050.013.25

Brickell, K., Picchioni, F., Natarajan, N., Guermond, V., Parsons, L., Zanello, G., & Bateman, M. (2020). Compounding crises of social reproduction: Microfinance, over-indebtedness and the COVID-19 pandemic. *World Development, 136*, 1–4. DOI: 10.1016/j.worlddev.2020.105087 PMID: 32834384

Brunswick, S. (2025). *What's space got to do with it?: An interstellar guide to success*. SB Global LLC.

Brynjolfsson, E., & McAfee, A. (2014). *The second machine age: Work, progress, and prosperity in a time of brilliant technologies*. WW Norton & Company.

Bureau of Labor Statistics. (2024). *Industry and occupational employment projections overview and highlights, 2023–33*. U.S. Bureau of Labor Statistics. https://fraser.stlouisfed.org/files/docs/publications/bls_mlr/bls_mlr_20241107.pdf

Burleigh, S. C., De Cola, T., Morosi, S., Jayousi, S., Cianca, E., & Fuchs, C. (2019). From Connectivity to advanced internet services: A comprehensive review of small satellites communications and networks. *Wireless Communications and Mobile Computing, 2019*(1), 6243505. DOI: 10.1155/2019/6243505

Çağman, N., & Enginoğlu, S. (2010). Soft matrix theory and its decision making. *Computers & Mathematics with Applications (Oxford, England), 59*(10), 3308–3314. DOI: 10.1016/j.camwa.2010.03.015

Çağman, N., & Enginoğlu, S. (2011). Fuzzy soft set theory and its applications. *Iranian Journal of Fuzzy Systems, 8*(3), 137–147. DOI: 10.22111/ijfs.2011.292

Castells, M. (2010). *The rise of the network society* (2nd ed.). John Wiley & Sons.

Cerioli, A., & Zani, S. (1990). A fuzzy approach to the measurement of poverty. In Dagum, C., & Zenga, M. (Eds.), *Income and wealth distribution, inequality and poverty* (pp. 272–284). Springer., DOI: 10.1007/978-3-642-84250-4_18

Charlesworth, A. (2012). Data protection, freedom of information and ethical review committees: Policies, practicalities and dilemmas. *Information Communication and Society, 15*(1), 85–103. DOI: 10.1080/1369118X.2011.637572

Charmes, K. (1999). *Informal sector, poverty and gender: A review of empirical evidence. Background Paper for the World Development Report 2001*. World Bank.

Cheli, B., & Lemmi, A. (1995). A totally fuzzy and relative approach to the multidimensional analysis of poverty. *Economic Notes, 24*(1), 115–134.

Chen, R. (2019). *Policy and regulatory issues with digital businesses.* World Bank Policy Research Working Paper. https://lawscopelibrary.com/admin/articles/policy-and-regulatory-issues-with-digital-businesses/WPS8948.pdf

Citaristi, I. (2022). United nations office for outer space affairs—unoosa. In *The Europa Directory of International Organizations 2022* (pp. 247–248). Routledge. DOI: 10.4324/9781003292548-53

Constantin, D.-L. (2021). Addressing spatial justice at lower territorial levels. some insights from the central and east European countries' perspective. *Regional Science Inquiry, 13*(2), 315–326.

Couldry, N., & Mejias, U. A. (2019). Data colonialism: Rethinking big data's relation to the contemporary subject. *Television & New Media, 20*(4), 336–349. DOI: 10.1177/1527476418796632

Couldry, N., & Mejias, U. A. (2023). The decolonial turn in data and technology research: What is at stake and where is it heading? *Information Communication and Society, 26*(4), 786–802. DOI: 10.1080/1369118X.2021.1986102

D'Ambrosio, C., & Chakravarty, S. R. (2019). Polarization orderings of poverty and richness. *Review of Income and Wealth, 65*(4), 794–820. DOI: 10.1111/roiw.12385

Daradkeh, L. (2023). Refugee camp streets: Unraveling lived space and community empowerment-a comparative case study analysis (Doctoral dissertation, Cardiff University).

Dare, P. M., & Fraser, C. S. (2001). Cover: Mapping informal settlements using high resolution satellite imagery. *International Journal of Remote Sensing*, *22*(8), 1399–1401. DOI: 10.1080/01431160120654

De la Barrera, F., Reyes-Paecke, S., & Banzhaf, E. (2016). Indicators for green spaces in contrasting urban settings. *Ecological Indicators*, *62*, 212–219. DOI: 10.1016/j.ecolind.2015.10.027

De la Barrera, F., Reyes-Paecke, S., Harris, J., Bascuñán, D., & Farías, J. M. (2016). People's perception influences on the use of green spaces in socio-economically differentiated neighborhoods. *Urban Forestry & Urban Greening*, *20*, 254–264. DOI: 10.1016/j.ufug.2016.09.007

De Sherbinin, A. (2018). The role of geospatial information in achieving the Sustainable Development Goals. *Geo-Spatial Information Science*, *21*(1), 21–30. DOI: 10.1080/10095020.2018.1498683

De Siqueira, A. C. (2012). The 2020 World Bank education strategy: Nothing new, or the same old gospel. In *The World Bank and Education* (pp. 67–81). Brill. DOI: 10.1007/978-94-6091-903-9_5

Dercon, S. (2022). *Gambling on development. Why some countries win and others lose*. Hurst & Company.

Desai, S., Barik, D., Choudhuri, P., Chouhan, B., & Sharma, O. P., & TIWARI, D. K. (2025). Rethinking Social Safety Nets in a Changing Society: Evidence from the India Human Development Survey. *India Policy Forum : [Papers]. India Policy Forum. Conference*, *21*(1), 96–136.

Desai, S., & Dubey, A. (2012). Caste in 21st century India: Competing narratives. *Economic and Political Weekly*, *46*(11), 40. PMID: 22736803

Devereux, S., & Sabates-Wheeler, R. (2004). *Transformative social protection*. Institute of Development Studies. https://hdl.handle.net/20.500.12413/4071

Dikeç, M. (2001). Justice and the spatial imagination. *Environment & Planning A*, *33*(10), 1785–1805. DOI: 10.1068/a3467

Duflo, E. (2012). Women empowerment and economic development. *Journal of Economic Literature*, *50*(4), 1051–1079. DOI: 10.1257/jel.50.4.1051

Elbers, B., Lanjouw, J., & Lanjouw, P. (2003). Micro-level estimation of poverty and inequality. *Econometrica*, *71*(1), 355–364. DOI: 10.1111/1468-0262.00399

Elwood, S. (2006). Critical issues in participatory GIS: Deconstructions, reconstructions, and new research directions. *Transactions in GIS*, *10*(5), 693–708. DOI: 10.1111/j.1467-9671.2006.01023.x

Engstrom, R., Hersh, J. S., & Newhouse, D. (2017). Poverty from space: Using high-resolution satellite imagery for estimating economic well-being. *World Bank Policy Research Working Paper No. 8284*. DOI: 10.1596/1813-9450-8284

ESCAP, U. (2003). Study on the use of space technology applications for poverty alleviation: trends, strategies and policy frameworks.

Fang, M. L., Canham, S. L., Battersby, L., Sixsmith, J., Wada, M., & Sixsmith, A. (2019). Exploring privilege in the digital divide: Implications for theory, policy, and practice. *The Gerontologist*, *59*(1), e1–e15. DOI: 10.1093/geront/gny037 PMID: 29750241

FAO. (2022). *The state of food and agriculture 2022: Leveraging automation in agriculture for transforming agrifood systems*. Food and Agriculture Organization., DOI: 10.4060/cb9479en

Feenstra, T. E., Inklaar, R., & Timmer, M. P. (2015). The next generation of the Penn World Table. *The American Economic Review*, *105*(10), 3150–3182. DOI: 10.1257/aer.20130954

Feng, F., Jun, Y. B., Liu, X., & Li, L. (2010). An adjustable approach to fuzzy soft set based decision making. *Journal of Computational and Applied Mathematics*, *234*(1), 10–20. DOI: 10.1016/j.cam.2009.11.055

Feng, F., Li, Y., & Leoreanu-Fotea, V. (2011). Application of level soft sets in decision making based on interval-valued fuzzy soft sets. *Computers & Mathematics with Applications (Oxford, England)*, *60*(6), 1756–1767. DOI: 10.1016/j.camwa.2010.07.006

Fields, A. (2004). Dualism in the labor market: A perspective on the Lewis model after half a century. *Manchester School*, *72*(6), 724–735. DOI: 10.1111/j.1467-9957.2004.00432.x

Fields, G. S. (2012). *Working hard, working poor: A global journey*. Oxford University Press.

Floro, A. M. (1995). *Economic restructuring, gender and the allocation of time*. World Bank. DOI: 10.1016/0305-750X(95)00092-Q

Friedman, T. L. (2006). *The world is flat: A brief history of the twenty-first century*. Farrar, Straus and Giroux.

Frolic, A. N., & Drolet, K. (2013). Ethics policy review: A case study in quality improvement. *Journal of Medical Ethics*, *39*(2), 98–103. DOI: 10.1136/medethics-2011-100461 PMID: 23118469

Gallegati, M., Palestrini, A., Petrini, M., & Russo, A. (2020). An agent-based model with network effects on income inequality. *Journal of Economic Interaction and Coordination*, *15*(4), 863–889. DOI: 10.1007/s11403-020-00295-y

Getzel, B., & Humphrey, C. (2024). *Streamlining MDB environmental, social and procurement safeguards while maintaining high standards*. ODI.

Ghadiri, M., Shabanpour, M., Hoseini, S. M., & Askarizadeh, D. (2018). Evaluation of fuzzy analytic hierarchy process in land suitability analysis for crop cultivation (case study: Takestan-Avaj plain). *Applied Ecology and Environmental Research*, *16*(2), 1667–1684. DOI: 10.15666/aeer/1602_16671684

Gibson, J., Olivia, S., & Rozelle, G. (2010). How deep is the depth of poverty? Evidence from Indonesia. *Asian Economic Journal*, *24*(2), 99–123. DOI: 10.1111/j.1467-8381.2010.02031.x

Gong, C., Chen, J., & Yu, S. (2013). Biotic homogenization of flora in urban green spaces. *Landscape and Urban Planning*, *120*, 158–169. DOI: 10.1016/j.landurbplan.2013.08.006

Greene, M. (2018). Socio-technical transitions and dynamics in everyday consumption practice. *Global Environmental Change*, *52*, 1–9. DOI: 10.1016/j.gloenvcha.2018.05.007

Hallegatte, S., Vogt-Schilb, A., Rozenberg, J., Bangalore, M., & Beaudet, C. (2020). From poverty to disaster and back: A review of the literature. *Economics of Disasters and Climate Change*, *4*(1), 223–247. DOI: 10.1007/s41885-020-00060-5

Han, B. H., & Wei, M. (2011). Distance and similarity measures for hesitant fuzzy soft sets. *Journal of Information and Computational Science*, *6*(4), 265–274.

Hargittai, E. (2003). The digital divide and what to do about it. In Jones, D. C. (Ed.), *New economy handbook* (Vol. 2003, pp. 821–839). Academic Press.

Harvey, D. (2010). *Social justice and the city* (Vol. 1). University of Georgia press. DOI: 10.1515/9780820336046

Heeks, R. (2008). ICT4D 2.0: The next phase of applying ICT for international development. *Computer*, *41*(6), 26–33. DOI: 10.1109/MC.2008.192

Helsper, E. (2021). *The digital disconnect: The social causes and consequences of digital inequalities*. SAGE Publications. DOI: 10.4135/9781526492982

Henderson, J. V., Storeygard, A., & Weil, D. N. (2012). Measuring economic growth from outer space. *The American Economic Review*, *102*(2), 994–1028. DOI: 10.1257/aer.102.2.994 PMID: 25067841

Hickel, J. (2017). *The divide: A brief guide to global inequality and its solutions*. Random House.

Hou, W., & Walz, U. (2014). Extraction of small biotopes using multi-temporal remote sensing. *International Journal of Remote Sensing*, *35*, 7245–7262. DOI: 10.1080/01431161.2014.967890

ILO. (2022). *World employment and social outlook: Trends 2022*. International Labour Organization., DOI: 10.54394/DSPL5113

International Labour Organization. (1998). *Women in the informal sector and their access to microfinance*. International Labour Organization.

International Labour Organization. (2012). *Rethinking economic growth: Towards productive and inclusive Arab societies*. International Labour Organization.

International Labour Organization. (2018). *Decent work and the Sustainable Development Goals: A guidebook on SDG labour market indicators*. International Labour Organization.

International Labour Organization. (2019). *Global wage report 2018/19: What lies behind gender pay gaps*. International Labour Organization.

International Labour Organization. (2020). *World employment and social outlook: Trends 2020*. International Labour Organization.

ISO. (2014). ISO 37120: Sustainable development of communities – Indicators for city services and quality of life.

Israel, E., & Frenkel, A. (2018). Social justice and spatial inequality: Toward a conceptual framework. *Progress in Human Geography*, *42*(5), 647–665. DOI: 10.1177/0309132517702969

ITU. (2023). *Measuring digital development: Facts and figures 2023*. International Telecommunication Union. https://www.itu.int/hub/publication/d-ind-ict_mdd-2023-1/

Iveson, K. (2011). Social or spatial justice? Marcuse and Soja on the right to the city. *City (London, England)*, *15*(2), 250–259. DOI: 10.1080/13604813.2011.568723

Jacobs, M. N., Colacci, A., Corvi, R., Vaccari, M., Aguila, M. C., Corvaro, M., Delrue, N., Desaulniers, D., Ertych, N., Jacobs, A., Luijten, M., Madia, F., Nishikawa, A., Ogawa, K., Ohmori, K., Paparella, M., Sharma, A. K., & Vasseur, P. (2020). Chemical carcinogen safety testing: OECD expert group international consensus on the development of an integrated approach for the testing and assessment of chemical non-genotoxic carcinogens. *Archives of Toxicology*, *94*(8), 2899–2923. DOI: 10.1007/s00204-020-02784-5 PMID: 32594184

Jana, C., & Pal, M. (2019). Multi-criteria decision making process based on some single-valued neutrosophic Dombi power aggregation operators. *Soft Computing*, *25*(7), 5055–5072. DOI: 10.1007/s00500-020-05509-z

Jean, N., Burke, M., Xie, M., Davis, W. M. A., Lobell, D. B., & Ermon, S. (2016). Combining satellite imagery and machine learning to predict poverty. *Science*, *353*(6301), 790–794. DOI: 10.1126/science.aaf7894 PMID: 27540167

Jerven, A. (2013). *Poor numbers: How we are misled by African development statistics and what to do about it*. Cornell University Press.

Johnson-Freese, J. (2016). *Space warfare in the 21st century: Arming the heavens*. Routledge., DOI: 10.4324/9781315529172

Jolliffe, D., & Prydz, E. B. (2021). Societal poverty: A relative and relevant measure. *The World Bank Economic Review*, *35*(1), 180–206. DOI: 10.1093/wber/lhz018

Joumard, I., & Kongsrud, P. M. (2003). Fiscal relations across government levels. *OECD Economic Studies*, *2003*(1), 155–229.

Jumayev, O. M. (2024). PHILOSOPHICAL ISSUES OF POVERTY: SOCIAL, ETHICAL, AND POLITICAL ASPECTS. European science review, (5-6), 41-49.

Kabeer, A. (2003). *Gender mainstreaming in poverty eradication and the Millennium Development Goals: A handbook for policy-makers and other stakeholders*. Commonwealth Secretariat. DOI: 10.14217/9781848598133-en

Kabeer, N. (2015). Gender, poverty, and inequality: A brief history of feminist contributions in the field of international development. *Gender and Development, 23*(2), 189–205. DOI: 10.1080/13552074.2015.1062300

Kanbur, R., & Venables, A. J. (2005). Spatial inequality and development. In Kanbur, R., & Venables, A. J. (Eds.), *Spatial inequality and development* (pp. 1–17). OUP Oxford. DOI: 10.1093/0199278636.003.0001

Karteris, M., Theodoridou, I., & Mallinis, G. (2016). Green roofs assessment using GIS and remote sensing. *Renewable & Sustainable Energy Reviews, 58*, 510–525. DOI: 10.1016/j.rser.2015.11.098

Kerle, N. (2013). Remote sensing based post-disaster damage mapping with collaborative methods. In Zlatanova, S., Peters, R., Dilo, A., & Scholten, H. (Eds.), *Intelligent systems for crisis management: Geo-information for disaster management (Gi4DM) 2012* (pp. 121–133). Springer., DOI: 10.1007/978-3-642-33218-0_9

Kganyago, M., & Mhangara, P. (2019). The role of African emerging space agencies in earth observation capacity building for facilitating the implementation and monitoring of the African development agenda: The case of African earth observation program. *ISPRS International Journal of Geo-Information, 8*(7), 292. DOI: 10.3390/ijgi8070292

Kim, Y., Safikhani, A., & Tepe, E. (2022). Machine learning application to spatio-temporal modeling of urban growth. *Computers, Environment and Urban Systems, 94*, 1–12. DOI: 10.1016/j.compenvurbsys.2022.101801

Kitchin, R. (2021). *The Data Revolution: A critical analysis of big data, open data and data infrastructures.*

Kitchin, R. (2014). Big Data, new epistemologies and paradigm shifts. *Big Data & Society, 1*(1), 2053951714528481. DOI: 10.1177/2053951714528481

Klir, G. J., & Yuan, B. (1995). *Fuzzy Sets and Fuzzy Logic*. Prentice Hall.

Koellner, E. (2024). Navigating the New Frontier: Ethical, Societal, and Legal Challenges in the Space Economy. In *AIAA Aviation Forum and Ascend 2024* (p. 4825).

Kong, Z., Gao, L., & Wang, L. (2014). An efficient decision making approach in incomplete soft set. *Applied Mathematical Modelling*, *38*(7-8), 2141–2150. DOI: 10.1016/j.apm.2013.10.009

Kornbluh, F. (2009). Is Work the Only Thing that Pays? The Guaranteed Income and Other Alternative Anti-Poverty Policies in Historical Perspective. *Nw. JL & Soc. Pol'y*, *4*, 61.

Krige, J. (2019). *How knowledge moves: Writing the transnational history of science and technology*. University of Chicago Press. DOI: 10.7208/chicago/9780226606040.001.0001

La Rosa, D. (2014). Accessibility to greenspaces: GIS-based indicators for sustainable planning. *Ecological Indicators*, *42*, 122–134. DOI: 10.1016/j.ecolind.2013.11.011

Lohmann, H., & Marx, I. (2018). *Handbook on in-work poverty*. Edward Elgar Publishing., DOI: 10.4337/9781784715632

Lopez, R. A., & Valdes, A. (1997). Fighting rural poverty in Latin America: New evidence of the effects of education, demographics, and access to land. *Economic Development and Cultural Change*, *45*(3), 487–507. DOI: 10.1086/452288

Lund, F., & Nicholson, J. (2003). *Chains of production, ladders of protection: Social protection for workers in the informal economy*. School of Development Studies.

Maas, J., van Dillen, S. M. E., Verheij, R. A., & Groenewegen, P. P. (2009). Social contacts as a mechanism between green space and health. *Health & Place*, *15*(2), 586–595. DOI: 10.1016/j.healthplace.2008.09.006 PMID: 19022699

Macdonald, G., & Macdonald, K. (2010). Safeguarding: A case for intelligent risk management. *British Journal of Social Work, 40*(4), 1174–1191. DOI: 10.1093/bjsw/bcq041

Maîtrot, M., Wood, G., & Devine, J. (2021). Understanding resilience: Lessons from lived experiences of extreme poverty in Bangladesh. *Development Policy Review, 39*(6), 894–910. DOI: 10.1111/dpr.12543

Maji, P. K., Biswas, R., & Roy, A. R. (2001). Fuzzy soft sets. *Journal of Fuzzy Mathematics, 9*(3), 589–602.

Maji, P. K., Biswas, R., & Roy, A. R. (2003). Soft set theory. *Computers & Mathematics with Applications (Oxford, England), 45*(4-5), 555–562. DOI: 10.1016/S0898-1221(03)00016-6

Maral, G., & Bousquet, M. (2020). *Satellite communications systems-Systems, techniques and technology* (6th ed.). Wiley. DOI: 10.1002/9781119673811

McGee, R., & Edwards, D. (2016). Introduction: Opening governance: Change, continuity and conceptual ambiguity. In D. Edwards & R. McGee (Eds.), *Opening governance* (Vol. 47, pp. 1–22). IDS Bulletin DOI: 10.19088/1968-2016.103

Mellor, J. W. (2017). *Agricultural development and economic transformation: Promoting growth with poverty reduction.* Palgrave Macmillan. DOI: 10.1007/978-3-319-65259-7

Melnyk, L. H., Sommer, H., Kubatko, O. V., Rabe, M., & Fedyna, S. M. (2020). *The economic and social drivers of renewable energy development in OECD countries.*

Meroni, M., Rembold, F., Urbano, F., Csak, G., Lemoine, G., Kerdiles, H., & Perez-Hoyos, A. (2019). *The warning classification scheme of ASAP–anomaly hot spots of agricultural production, v8. 0.* Luxembourg: Publications Office of the European Union. https://data.europa.eu/doi/10.2760/6814458

Milanovic, B. (2019). *Capitalism, alone: The future of the system that rules the world*. Harvard University Press.

Modise, J. M. (2025). ERADICATING POVERTY IN ALL ITS DIMENSIONS: A POLICY-ORIENTED APPROACH TO ACHIEVING SUSTAINABLE DEVELOPMENT GOAL.

Molodtsov, D. (1999). Soft set theory—First results. *Computers & Mathematics with Applications (Oxford, England)*, *37*(4-5), 19–31. DOI: 10.1016/S0898-1221(99)00056-5

Monico, J. F. G. (2004). United Nations office for outer space affairs GNSS web-pages. *GPS Solutions*, *8*(2), 112–114. DOI: 10.1007/s10291-004-0096-4

Moravec, E. T. (2024). Digital Divide and Social Inclusion: Strategies to Enhance Digital Participation of Marginalized Groups. *Future-Social Sciences and Digital Society*, *1*(2).

Mulhaney, A., Parkinson, M., & Abdulfatah, A. K. (2024). A Human-Centric Approach to Digital Transformation. In A. Visvizi, O. Troisi, V. Corvello, & M. Grimaldi (Eds.), *Research and Innovation Forum 2024* (pp. 621–634). Springer. DOI: 10.1007/978-3-031-78623-5_49

Nardo, M., Saisana, M., Saltelli, A., Tarantola, S., Hoffman, A., & Giovannini, E. (2019). *Handbook on constructing composite indicators: Methodology and user guide*. OECD Publishing., DOI: 10.1787/9789264043466-

Nguyen, H. T., & Kreinovich, V. (1997). *Applications of Continuous Mathematics to Computer Science*. Kluwer. DOI: 10.1007/978-94-017-0743-5

Nguyen, H. T., & Walker, E. A. (2006). *First Course in Fuzzy Logic*. CRC Press.

Niemelä, J. (2014). Ecology of urban green spaces: Key research questions. *Landscape and Urban Planning*, *125*, 298–303. DOI: 10.1016/j.landurbplan.2013.07.014

O'Neil, I., & Ucbasaran, D. (2016). Balancing "what matters to me" with "what matters to them": Exploring the legitimation process of environmental entrepreneurs. *Journal of Business Venturing*, *31*(2), 133–152. DOI: 10.1016/j.jbusvent.2015.12.001

Organisation for Economic Co-operation and Development. (2019). *Society at a glance 2019: OECD social indicators*. OECD Publishing.

Peng, X., & Garg, H. (2018). Algorithms for interval-valued fuzzy soft sets in emergency decision making based on WDBA and CODAS with new information measure. *Computers & Industrial Engineering*, *119*, 439–452. DOI: 10.1016/j.cie.2018.04.001 PMID: 32288046

Phillips, N. (2017). Power and inequality in the global political economy. *International Affairs*, *93*(2), 429–444. DOI: 10.1093/ia/iix019

Purtova, N., & van Maanen, G. (2024). Data as an economic good, data as a commons, and data governance. *Law, Innovation and Technology*, *16*(1), 1–42. DOI: 10.1080/17579961.2023.2265270 PMID: 39634789

Qin, H., Ma, X., Herawan, T., & Zain, J. M. (2020). A novel soft set approach to decision making. *Knowledge-Based Systems*, *152*, 116–125. DOI: 10.1016/j.knosys.2018.04.018

Ravallion, M. (2016). Poverty lines across the world. *The World Bank Research Observer*, *31*(2), 259–287. DOI: 10.1093/wbro/lkw003

Ravallion, M. (2016). *The economics of poverty: History, measurement, and policy*. Oxford University Press. DOI: 10.1093/acprof:oso/9780190212766.001.0001

Rembold, F., Meroni, M., Urbano, F., Csak, G., Kerdiles, H., Perez-Hoyos, A., Lemoine, G., Leo, O., & Negre, T. (2019). ASAP: A new global early warning system to detect anomaly hot spots of agricultural production for food security analysis. *Agricultural Systems*, *168*, 247–257. DOI: 10.1016/j.agsy.2018.07.002 PMID: 30774185

Rodgers, G. (2018). Inequality in the Indian growth regime. *Indian Journal of Human Development*, *12*(2), 134–148. DOI: 10.1177/0973703018793956

Rodolfa, K. T., Saleiro, P., & Ghani, R. (2020). Bias and fairness. In *Big data and social science* (pp. 281–312). Chapman and Hall/CRC. DOI: 10.1201/9780429324383-11

Romero, H., Vásquez, A., Fuentes, C., Salgado, M., Schmidt, A., & Banzhaf, E. (2012). Assessing urban environmental segregation (UES). The case of Santiago de Chile. *Ecological Indicators*, *23*, 76–87. DOI: 10.1016/j.ecolind.2012.03.012

Rostow, W. W. (1960). *The stages of economic growth: A non-communist manifesto*. Cambridge University Press.

Rostow, W. W. (2000). The stages of economic growth. In Sanderson, S. K. (Ed.), *Sociological worlds: Comparative and historical readings on society* (pp. 130–134). Routledge., DOI: 10.4324/9781315063362-14

Roy, A. R., & Maji, P. K. (2007). A fuzzy soft set theoretic approach to decision making problems. *Journal of Computational and Applied Mathematics*, *203*(2), 412–418. DOI: 10.1016/j.cam.2006.04.008

Ruighaver, A. B., Maynard, S. B., & Warren, M. (2010). Ethical decision making: Improving the quality of acceptable use policies. *Computers & Security*, *29*(7), 731–736. DOI: 10.1016/j.cose.2010.05.004

Sachs, J. D. (2015). *The age of sustainable development*. Columbia University Press. DOI: 10.7312/sach17314

Sandau, R. (2010). Status and trends of small satellite missions for Earth observation. *acta astronautica, 66*(1-2), 1-12. DOI: 10.1016/j.actaastro.2009.06.008

Santos, M. E., & Villatoro, P. (2018). A multidimensional poverty index for Latin America. *Review of Income and Wealth*, *64*(1), 52–82. DOI: 10.1111/roiw.12275

Sayed, Z. (2015). Transforming our world: The 2030 agenda for sustainable development. *Ethics & Critical Thinking Journal*, *2015*(3), 112–142.

Schipperijn, J., Stigsdotter, U. K., Randrup, T. B., & Troelsen, J. (2010). Influences on the use of green space. *Urban Forestry & Urban Greening*, *9*(1), 25–32. DOI: 10.1016/j.ufug.2009.09.002

Schwab, K. (2016). *The Fourth Industrial Revolution*. World Economic Forum.

Sen, A. (2015). Development as freedom (1999). In Roberts, J. T., Hite, A. B., & Chorev, N. (Eds.), *The globalization and development reader: Perspectives on development and global change* (2nd ed., pp. 525–548). Wiley Blackwell.

Smith, D. J., & Ward, M. W. (2014). Mapping poverty and inequality: New perspectives using GIS and remote sensing. *Computers, Environment and Urban Systems*, *42*, 35–45. DOI: 10.1016/j.compenvurbsys.2013.10.001

Soja, E. (2010). Spatializing the urban, Part I. *City (London, England)*, *14*(6), 629–635. DOI: 10.1080/13604813.2010.539371

Squire, L. M. (1996). Poverty in developing countries: Trends and policy responses. *The World Bank Research Observer*, *11*(2), 153–171. DOI: 10.1093/wbro/11.2.153

Stankutė, S., & Asche, H. (2009, June). An integrative approach to geospatial data fusion. In *International Conference on Computational Science and Its Applications* (pp. 490-504). Berlin, Heidelberg: Springer Berlin Heidelberg. DOI: 10.1007/978-3-642-02454-2_35

Steele, J. E., Sundsøy, P. R., Pezzulo, C., Alegana, V. A., Bird, T. J., Blumenstock, J., Bjelland, J., Engø-Monsen, K., de Montjoye, Y.-A., Iqbal, A. M., Hadiuzzaman, K. N., Lu, X., Wetter, E., Tatem, A. J., & Bengtsson, L. (2017). Mapping poverty using mobile phone and satellite data. *Journal of the Royal Society, Interface*, *14*(127), 20160690. DOI: 10.1098/rsif.2016.0690 PMID: 28148765

Sumner, A. (2016). *Global poverty: Deprivation, distribution, and development since the cold War*. Oxford University Press. DOI: 10.1093/acprof:oso/9780198703525.001.0001

Tatem, A., Campbell, C., & Guerra-Arias, D. (2014). Mapping for maternal and newborn health: The value of spatial data in health planning. *International Journal of Health Geographics*, *13*(1), 1–9. DOI: 10.1186/1476-072X-13-13

Taylor, B. T., Fernando, P., Bauman, A. E., Williamson, A., Craig, J. C., & Redman, S. (2011). Measuring the quality of public open space using Google Earth. *American Journal of Preventive Medicine*, *40*(2), 105–112. DOI: 10.1016/j.amepre.2010.10.024 PMID: 21238857

Tennin, K. L., & Brunswick, S. (2024). Sustainable approaches to poverty and challenges from the 21st century. In Goel, R., Singh, T., Rahman, M. M., Islam, Q. T., & Baral, S. K. (Eds.), *Understanding the multi-dimensional nature of poverty* (pp. 13–41). Emerald Publishing., DOI: 10.1108/978-1-83753-292-620241002

Thornton, P. K., Ericksen, P. J., Herrero, M., & Challinor, A. J. (2014). Climate variability and vulnerability to climate change: A review. *Global Change Biology*, *20*(11), 3313–3328. DOI: 10.1111/gcb.12581 PMID: 24668802

Tian, Y., Jim, C. Y., & Wang, H. (2014). Assessing ecological quality of urban green spaces. *Landscape and Urban Planning*, *121*, 97–108. DOI: 10.1016/j.landurbplan.2013.10.001

Tokman, M. (1989). Integrating the informal sector into the modernization process. *World Development*, *17*(7), 1065–1075. DOI: 10.1016/0305-750X(89)90113-1

Tokman, M. (1991). *Flexible work in Latin America*. International Labour Organization.

Toyama, K. (2015). *Geek heresy: Rescuing social change from the cult of technology*. PublicAffairs.

U.S. Census Bureau. (2023). *Census bureau releases new report about persistent poverty at county and census-tract level*. United States Census Bureau. https://www.census.gov/newsroom/press-releases/2023/persistent-poverty.html

UNDRR. (2023). *Global assessment report on disaster risk reduction 2023: Mapping resilience for the sustainable development goals*. Geneva: United Nations Office for Disaster Risk Reduction. https://digitallibrary.un.org/record/4022485?ln=en&v=pdf

UNESCO. (2023). *Global education monitoring report 2023: Technology in education: A tool on whose terms?* UNESCO. https://www.unesco.org/gem-report/en/technology

United Nations Development Programme. (2019). *Human development report 2019: Beyond income, beyond averages, beyond today—inequalities in human development in the 21st century*. UNDP.

United Nations Development Programme. (2020). *Human development report 2020: The next frontier—Human development and the Anthropocene*. UNDP.

United Nations Economic and Social Commission for Western Asia. (2014). *Arab middle class: Measurement and role in driving change*. UN ESCWA.

United Nations. (2015). *Transforming our world: The 2030 agenda for sustainable development*. United Nations. https://sdgs.un.org/2030agenda

UNOOSA. (2024). *From strategy to action: Annual report 2024*. Vienna: United Nations Office for Outer Space Affairs. https://www.unoosa.org/documents/pdf/annualreport/UNOOSA_Annual_Report_2024.pdf

Uwayezu, E., & de Vries, W. T. (2019). Scoping land tenure security for the poor and low-income urban dwellers from a spatial justice lens. *Habitat International*, *91*, 102016. DOI: 10.1016/j.habitatint.2019.102016

Van der Linden, M. (2019). The International Labour Organization, 1919–2019: An Appraisal. *Labor*, *16*(2), 11–41. DOI: 10.1215/15476715-7323601

Van Dijk, J. (2020). *The digital divide*. John Wiley & Sons.

Van Herzele, A., & Wiedemann, T. (2003). A monitoring tool for the provision of accessible and attractive urban green spaces. *Landscape and Urban Planning*, *63*(2), 109–126. DOI: 10.1016/S0169-2046(02)00192-5

Van Laar, E., Van Deursen, A. J. A. M., Van Dijk, J. A. G. M., & De Haan, J. (2020). Determinants of 21st-century skills and 21st-century digital skills for workers: A systematic literature review. *SAGE Open*, *10*(1), 2158244019900176. DOI: 10.1177/2158244019900176

Villagran de Leon, J. C. (2023). An improved understanding of natural hazards and disasters through the use of satellite technologies: Contributions from un-spider. *Sustainability (Basel)*, *15*(13), 10624. DOI: 10.3390/su151310624

Viso, M. (2023). ISRO, India. In *Encyclopedia of Astrobiology* (pp. 1558–1559). Springer. DOI: 10.1007/978-3-662-65093-6_1784

Wagle, U. (2002). Rethinking poverty: Definition and measurement. *International Social Science Journal*, *54*(171), 155–165. DOI: 10.1111/1468-2451.00366

Warschauer, M. (2003). *Technology and social inclusion: Rethinking the digital divide*. MIT Press. DOI: 10.7551/mitpress/6699.001.0001

Weeks, J. R., Getis, A., Hill, A. G., Gadalla, M. S., & Rashed, T. (2020). The fertility transition in Egypt: Intraurban patterns in Cairo. *Annals of the Association of American Geographers*, *94*(1), 74–93. DOI: 10.1111/j.1467-8306.2004.09401005.x

WHO. (2025). *Global strategy on digital health 2020-2027*. Geneva: World Health Organization. https://iris.who.int/server/api/core/bitstreams/a64d584f-015b-471f-9d86-9ea263cfb516/content

Worl Bank. (2022). *Poverty and shared prosperity 2022: Correcting course*. Washington DC: World Bank. https://openknowledge.worldbank.org/server/api/core/bitstreams/b96b361a-a806-5567-8e8a-b14392e11fa0/content

World Bank. (2013). *World development report 2013: Jobs*. World Bank.

World Bank. (2021). *Informality: Causes, consequences and policy responses*. World Bank.

Wright-Wendel, H., Zarger, R. K., & Mihelcic, J. R. (2012). Accessibility and usability: Green space perceptions in Latin America. *Landscape and Urban Planning*, *107*, 272–282. DOI: 10.1016/j.landurbplan.2012.06.003

Xaxa, V. (2011). The status of tribal children in India: A historical perspective. *IHD-UNICEF Working Paper, 7.*

Yao, L., Liu, J., Wang, R., Yin, K., & Han, B. (2014). Effective green equivalent—A measure of public green spaces for cities. *Ecological Indicators*, *47*, 123–127. DOI: 10.1016/j.ecolind.2014.07.009

Zadeh, L. A. (1965). Fuzzy sets. *Information and Control*, *8*(3), 338–353. DOI: 10.1016/S0019-9958(65)90241-X

Zhang, H., Xiong, L., & Ma, W. (2020). On interval-valued hesitant fuzzy soft sets. *Mathematical Problems in Engineering*, *2020*, 1649360. Advance online publication. DOI: 10.1155/2020/1649360

Zhou, X., & Kim, J. (2013). Social disparities in tree canopy and park accessibility. *Urban Forestry & Urban Greening*, *12*(1), 88–97. DOI: 10.1016/j.ufug.2012.11.004

Related References

To continue our tradition of advancing academic research, we have compiled a list of recommended IGI Global readings. These references will provide additional information and guidance to further enrich your knowledge and assist you with your own research and future publications.

Abbasnejad, B., Moeinzadeh, S., Ahankoob, A., & Wong, P. S. (2021). The Role of Collaboration in the Implementation of BIM-Enabled Projects. In Underwood, J., & Shelbourn, M. (Eds.), *Handbook of Research on Driving Transformational Change in the Digital Built Environment* (pp. 27–62). IGI Global. https://doi.org/10.4018/978-1-7998-6600-8.ch002

Abdulrahman, K. O., Mahamood, R. M., & Akinlabi, E. T. (2022). Additive Manufacturing (AM): Processing Technique for Lightweight Alloys and Composite Material. In K. Kumar, B. Babu, & J. Davim (Ed.), *Handbook of Research on Advancements in the Processing, Characterization, and Application of Lightweight Materials* (pp. 27-48). IGI Global. https://doi.org/10.4018/978-1-7998-7864-3.ch002

Agrawal, R., Sharma, P., & Saxena, A. (2021). A Diamond Cut Leather Substrate Antenna for BAN (Body Area Network) Application. In Singh, V., Dubey, V., Saxena, A., Tiwari, R., & Sharma, H. (Eds.), *Emerging Materials and Advanced Designs for Wearable Antennas* (pp. 54–59). IGI Global. https://doi.org/10.4018/978-1-7998-7611-3.ch004

Ahmad, F., Al-Ammar, E. A., & Alsaidan, I. (2022). Battery Swapping Station: A Potential Solution to Address the Limitations of EV Charging Infrastructure. In Alam, M., Pillai, R., & Murugesan, N. (Eds.), *Developing Charging Infrastructure and Technologies for Electric Vehicles* (pp. 195–207). IGI Global. DOI: 10.4018/978-1-7998-6858-3.ch010

Aikhuele, D. (2018). A Study of Product Development Engineering and Design Reliability Concerns. *International Journal of Applied Industrial Engineering*, *5*(1), 79–89. DOI: 10.4018/IJAIE.2018010105

Al-Khatri, H., & Al-Atrash, F. (2021). Occupants' Habits and Natural Ventilation in a Hot Arid Climate. In González-Lezcano, R. (Ed.), *Advancements in Sustainable Architecture and Energy Efficiency* (pp. 146–168). IGI Global. https://doi.org/10.4018/978-1-7998-7023-4.ch007

Al-Shebeeb, O. A., Rangaswamy, S., Gopalakrishan, B., & Devaru, D. G. (2017). Evaluation and Indexing of Process Plans Based on Electrical Demand and Energy Consumption. *International Journal of Manufacturing, Materials, and Mechanical Engineering*, *7*(3), 1–19. DOI: 10.4018/IJMMME.2017070101

Amuda, M. O., Lawal, T. F., & Akinlabi, E. T. (2017). Research Progress on Rheological Behavior of AA7075 Aluminum Alloy During Hot Deformation. *International Journal of Materials Forming and Machining Processes*, *4*(1), 53–96. DOI: 10.4018/IJMFMP.2017010104

Amuda, M. O., Lawal, T. F., & Mridha, S. (2021). Microstructure and Mechanical Properties of Silicon Carbide-Treated Ferritic Stainless Steel Welds. In Burstein, L. (Ed.), *Handbook of Research on Advancements in Manufacturing, Materials, and Mechanical Engineering* (pp. 395–411). IGI Global. https://doi.org/10.4018/978-1-7998-4939-1.ch019

Anikeev, V., Gasem, K. A., & Fan, M. (2021). Application of Supercritical Technologies in Clean Energy Production: A Review. In Chen, L. (Ed.), *Handbook of Research on Advancements in Supercritical Fluids Applications for Sustainable Energy Systems* (pp. 792–821). IGI Global. https://doi.org/10.4018/978-1-7998-5796-9.ch022

Arafat, M. Y., Saleem, I., & Devi, T. P. (2022). Drivers of EV Charging Infrastructure Entrepreneurship in India. In Alam, M., Pillai, R., & Murugesan, N. (Eds.), *Developing Charging Infrastructure and Technologies for Electric Vehicles* (pp. 208–219). IGI Global. https://doi.org/10.4018/978-1-7998-6858-3.ch011

Araujo, A., & Manninen, H. (2022). Contribution of Project-Based Learning on Social Skills Development: An Industrial Engineer Perspective. In Alves, A., & van Hattum-Janssen, N. (Eds.), *Training Engineering Students for Modern Technological Advancement* (pp. 119–145). IGI Global. https://doi.org/10.4018/978-1-7998-8816-1.ch006

Armutlu, H. (2018). Intelligent Biomedical Engineering Operations by Cloud Computing Technologies. In Kose, U., Guraksin, G., & Deperlioglu, O. (Eds.), *Nature-Inspired Intelligent Techniques for Solving Biomedical Engineering Problems* (pp. 297–317). Hershey, PA: IGI Global. DOI: 10.4018/978-1-5225-4769-3.ch015

Atik, M., Sadek, M., & Shahrour, I. (2017). Single-Run Adaptive Pushover Procedure for Shear Wall Structures. In Plevris, V., Kremmyda, G., & Fahjan, Y. (Eds.), *Performance-Based Seismic Design of Concrete Structures and Infrastructures* (pp. 59–83). Hershey, PA: IGI Global. DOI: 10.4018/978-1-5225-2089-4.ch003

Attia, H. (2021). Smart Power Microgrid Impact on Sustainable Building. In González-Lezcano, R. (Ed.), *Advancements in Sustainable Architecture and Energy Efficiency* (pp. 169–194). IGI Global. https://doi.org/10.4018/978-1-7998-7023-4.ch008

Aydin, A., Akyol, E., Gungor, M., Kaya, A., & Tasdelen, S. (2018). Geophysical Surveys in Engineering Geology Investigations With Field Examples. In Ceryan, N. (Ed.), *Handbook of Research on Trends and Digital Advances in Engineering Geology* (pp. 257–280). Hershey, PA: IGI Global. DOI: 10.4018/978-1-5225-2709-1.ch007

Ayoobkhan, M. U. D., Y., A., J., Easwaran, B., & R., T. (2021). Smart Connected Digital Products and IoT Platform With the Digital Twin. In P. Vasant, G. Weber, & W. Punurai (Ed.), *Research Advancements in Smart Technology, Optimization, and Renewable Energy* (pp. 330-350). IGI Global. https://doi.org/DOI: 10.4018/978-1-7998-3970-5.ch016

Baeza Moyano, D., & González Lezcano, R. A. (2021). The Importance of Light in Our Lives: Towards New Lighting in Schools. In González-Lezcano, R. (Ed.), *Advancements in Sustainable Architecture and Energy Efficiency* (pp. 239–256). IGI Global. https://doi.org/10.4018/978-1-7998-7023-4.ch011

Bagdadee, A. H. (2021). A Brief Assessment of the Energy Sector of Bangladesh. *International Journal of Energy Optimization and Engineering*, *10*(1), 36–55. DOI: 10.4018/IJEOE.2021010103

Baklezos, A. T., & Hadjigeorgiou, N. G. (2021). Magnetic Sensors for Space Applications and Magnetic Cleanliness Considerations. In Nikolopoulos, C. (Ed.), *Recent Trends on Electromagnetic Environmental Effects for Aeronautics and Space Applications* (pp. 147–185). IGI Global. https://doi.org/10.4018/978-1-7998-4879-0.ch006

Bas, T. G. (2017). Nutraceutical Industry with the Collaboration of Biotechnology and Nutrigenomics Engineering: The Significance of Intellectual Property in the Entrepreneurship and Scientific Research Ecosystems. In Bas, T., & Zhao, J. (Eds.), *Comparative Approaches to Biotechnology Development and Use in Developed and Emerging Nations* (pp. 1–17). Hershey, PA: IGI Global. DOI: 10.4018/978-1-5225-1040-6.ch001

Bazeer Ahamed, B., & Periakaruppan, S. (2021). Taxonomy of Influence Maximization Techniques in Unknown Social Networks. In P. Vasant, G. Weber, & W. Punurai (Eds.), *Research Advancements in Smart Technology, Optimization, and Renewable Energy* (pp. 351-363). IGI Global. https://doi.org/10.4018/978-1-7998-3970-5.ch017

Beale, R., & André, J. (2017). *Design Solutions and Innovations in Temporary Structures*. Hershey, PA: IGI Global. DOI: 10.4018/978-1-5225-2199-0

Behnam, B. (2017). Simulating Post-Earthquake Fire Loading in Conventional RC Structures. In Samui, P., Chakraborty, S., & Kim, D. (Eds.), *Modeling and Simulation Techniques in Structural Engineering* (pp. 425–444). Hershey, PA: IGI Global. DOI: 10.4018/978-1-5225-0588-4.ch015

Ben Hamida, I., Salah, S. B., Msahli, F., & Mimouni, M. F. (2018). Distribution Network Reconfiguration Using SPEA2 for Power Loss Minimization and Reliability Improvement. *International Journal of Energy Optimization and Engineering*, *7*(1), 50–65. DOI: 10.4018/IJEOE.2018010103

Bentarzi, H. (2021). Fault Tree-Based Root Cause Analysis Used to Study Mal-Operation of a Protective Relay in a Smart Grid. In Recioui, A., & Bentarzi, H. (Eds.), *Optimizing and Measuring Smart Grid Operation and Control* (pp. 289–308). IGI Global. https://doi.org/10.4018/978-1-7998-4027-5.ch012

Beysens, D. A., Garrabos, Y., & Zappoli, B. (2021). Thermal Effects in Near-Critical Fluids: Piston Effect and Related Phenomena. In Chen, L. (Ed.), *Handbook of Research on Advancements in Supercritical Fluids Applications for Sustainable Energy Systems* (pp. 1–31). IGI Global. https://doi.org/10.4018/978-1-7998-5796-9.ch001

Bhaskar, S. V., & Kudal, H. N. (2017). Effect of TiCN and AlCrN Coating on Tribological Behaviour of Plasma-nitrided AISI 4140 Steel. *International Journal of Surface Engineering and Interdisciplinary Materials Science*, *5*(2), 1–17. DOI: 10.4018/IJSEIMS.2017070101

Bhuyan, D. (2018). Designing of a Twin Tube Shock Absorber: A Study in Reverse Engineering. In Kumar, K., & Davim, J. (Eds.), *Design and Optimization of Mechanical Engineering Products* (pp. 83–104). Hershey, PA: IGI Global. DOI: 10.4018/978-1-5225-3401-3.ch005

Blumberg, G. (2021). Blockchains for Use in Construction and Engineering Projects. In Underwood, J., & Shelbourn, M. (Eds.), *Handbook of Research on Driving Transformational Change in the Digital Built Environment* (pp. 179–208). IGI Global. https://doi.org/10.4018/978-1-7998-6600-8.ch008

Bolboaca, A. M. (2021). Considerations Regarding the Use of Fuel Cells in Combined Heat and Power for Stationary Applications. In Badea, G., Felseghi, R., & A chilean, I. (Eds.), *Hydrogen Fuel Cell Technology for Stationary Applications* (pp. 239–275). IGI Global. https://doi.org/10.4018/978-1-7998-4945-2.ch010

Burstein, L. (2021). Simulation Tool for Cable Design. In Burstein, L. (Ed.), *Handbook of Research on Advancements in Manufacturing, Materials, and Mechanical Engineering* (pp. 54–74). IGI Global. https://doi.org/10.4018/978-1-7998-4939-1.ch003

Calderon, F. A., Giolo, E. G., Frau, C. D., Rengel, M. G., Rodriguez, H., Tornello, M., & Gallucci, R. (2018). Seismic Microzonation and Site Effects Detection Through Microtremors Measures: A Review. In Ceryan, N. (Ed.), *Handbook of Research on Trends and Digital Advances in Engineering Geology* (pp. 326–349). Hershey, PA: IGI Global. DOI: 10.4018/978-1-5225-2709-1.ch009

Ceryan, N., & Can, N. K. (2018). Prediction of The Uniaxial Compressive Strength of Rocks Materials. In Ceryan, N. (Ed.), *Handbook of Research on Trends and Digital Advances in Engineering Geology* (pp. 31–96). Hershey, PA: IGI Global. DOI: 10.4018/978-1-5225-2709-1.ch002

Ceryan, S. (2018). Weathering Indices Used in Evaluation of the Weathering State of Rock Material. In Ceryan, N. (Ed.), *Handbook of Research on Trends and Digital Advances in Engineering Geology* (pp. 132–186). Hershey, PA: IGI Global. DOI: 10.4018/978-1-5225-2709-1.ch004

Chen, H., Padilla, R. V., & Besarati, S. (2017). Supercritical Fluids and Their Applications in Power Generation. In Chen, L., & Iwamoto, Y. (Eds.), *Advanced Applications of Supercritical Fluids in Energy Systems* (pp. 369–402). Hershey, PA: IGI Global. DOI: 10.4018/978-1-5225-2047-4.ch012

Chen, H., Padilla, R. V., & Besarati, S. (2021). Supercritical Fluids and Their Applications in Power Generation. In Chen, L. (Ed.), *Handbook of Research on Advancements in Supercritical Fluids Applications for Sustainable Energy Systems* (pp. 566–599). IGI Global. https://doi.org/10.4018/978-1-7998-5796-9.ch016

Chen, L. (2017). Principles, Experiments, and Numerical Studies of Supercritical Fluid Natural Circulation System. In Chen, L., & Iwamoto, Y. (Eds.), *Advanced Applications of Supercritical Fluids in Energy Systems* (pp. 136–187). Hershey, PA: IGI Global. DOI: 10.4018/978-1-5225-2047-4.ch005

Chen, L. (2021). Principles, Experiments, and Numerical Studies of Supercritical Fluid Natural Circulation System. In Chen, L. (Ed.), *Handbook of Research on Advancements in Supercritical Fluids Applications for Sustainable Energy Systems* (pp. 219–269). IGI Global. https://doi.org/10.4018/978-1-7998-5796-9.ch007

Chiba, Y., Marif, Y., Henini, N., & Tlemcani, A. (2021). Modeling of Magnetic Refrigeration Device by Using Artificial Neural Networks Approach. *International Journal of Energy Optimization and Engineering*, *10*(4), 68–76. https://doi.org/10.4018/IJEOE.2021100105

Clementi, F., Di Sciascio, G., Di Sciascio, S., & Lenci, S. (2017). Influence of the Shear-Bending Interaction on the Global Capacity of Reinforced Concrete Frames: A Brief Overview of the New Perspectives. In Plevris, V., Kremmyda, G., & Fahjan, Y. (Eds.), *Performance-Based Seismic Design of Concrete Structures and Infrastructures* (pp. 84–111). Hershey, PA: IGI Global. DOI: 10.4018/978-1-5225-2089-4.ch004

Codinhoto, R., Fialho, B. C., Pinti, L., & Fabricio, M. M. (2021). BIM and IoT for Facilities Management: Understanding Key Maintenance Issues. In Underwood, J., & Shelbourn, M. (Eds.), *Handbook of Research on Driving Transformational Change in the Digital Built Environment* (pp. 209–231). IGI Global. DOI: 10.4018/978-1-7998-6600-8.ch009

Cortés-Polo, D., Calle-Cancho, J., Carmona-Murillo, J., & González-Sánchez, J. (2017). Future Trends in Mobile-Fixed Integration for Next Generation Networks: Classification and Analysis. *International Journal of Vehicular Telematics and Infotainment Systems*, *1*(1), 33–53. DOI: 10.4018/IJVTIS.2017010103

Costa, H. G., Sheremetieff, F. H., & Araújo, E. A. (2022). Influence of Game-Based Methods in Developing Engineering Competences. In Alves, A., & van Hattum-Janssen, N. (Eds.), *Training Engineering Students for Modern Technological Advancement* (pp. 69–88). IGI Global. https://doi.org/10.4018/978-1-7998-8816-1.ch004

Cui, X., Zeng, S., Li, Z., Zheng, Q., Yu, X., & Han, B. (2018). Advanced Composites for Civil Engineering Infrastructures. In Kumar, K., & Davim, J. (Eds.), *Composites and Advanced Materials for Industrial Applications* (pp. 212–248). Hershey, PA: IGI Global. DOI: 10.4018/978-1-5225-5216-1.ch010

Dalgıç, S., & Kuşku, İ. (2018). Geological and Geotechnical Investigations in Tunneling. In Ceryan, N. (Ed.), *Handbook of Research on Trends and Digital Advances in Engineering Geology* (pp. 482–529). Hershey, PA: IGI Global. DOI: 10.4018/978-1-5225-2709-1.ch014

Dang, C., & Hihara, E. (2021). Study on Cooling Heat Transfer of Supercritical Carbon Dioxide Applied to Transcritical Carbon Dioxide Heat Pump. In Chen, L. (Ed.), *Handbook of Research on Advancements in Supercritical Fluids Applications for Sustainable Energy Systems* (pp. 451–493). IGI Global. https://doi.org/10.4018/978-1-7998-5796-9.ch013

Daus, Y., Kharchenko, V., & Yudaev, I. (2021). Research of Solar Energy Potential of Photovoltaic Installations on Enclosing Structures of Buildings. *International Journal of Energy Optimization and Engineering*, *10*(4), 18–34. https://doi.org/10.4018/IJEOE.2021100102

Daus, Y., Kharchenko, V., & Yudaev, I. (2021). Optimizing Layout of Distributed Generation Sources of Power Supply System of Agricultural Object. *International Journal of Energy Optimization and Engineering*, *10*(3), 70–84. https://doi.org/10.4018/IJEOE.2021070104

de la Varga, D., Soto, M., Arias, C. A., van Oirschot, D., Kilian, R., Pascual, A., & Álvarez, J. A. (2017). Constructed Wetlands for Industrial Wastewater Treatment and Removal of Nutrients. In Val del Río, Á., Campos Gómez, J., & Mosquera Corral, A. (Eds.), *Technologies for the Treatment and Recovery of Nutrients from Industrial Wastewater* (pp. 202–230). Hershey, PA: IGI Global. DOI: 10.4018/978-1-5225-1037-6.ch008

Deb, S., Ammar, E. A., AlRajhi, H., Alsaidan, I., & Shariff, S. M. (2022). V2G Pilot Projects: Review and Lessons Learnt. In Alam, M., Pillai, R., & Murugesan, N. (Eds.), *Developing Charging Infrastructure and Technologies for Electric Vehicles* (pp. 252–267). IGI Global. https://doi.org/10.4018/978-1-7998-6858-3.ch014

Dekhandji, F. Z., & Rais, M. C. (2021). A Comparative Study of Power Quality Monitoring Using Various Techniques. In Recioui, A., & Bentarzi, H. (Eds.), *Optimizing and Measuring Smart Grid Operation and Control* (pp. 259–288). IGI Global. https://doi.org/10.4018/978-1-7998-4027-5.ch011

Deperlioglu, O. (2018). Intelligent Techniques Inspired by Nature and Used in Biomedical Engineering. In Kose, U., Guraksin, G., & Deperlioglu, O. (Eds.), *Nature-Inspired Intelligent Techniques for Solving Biomedical Engineering Problems* (pp. 51–77). Hershey, PA: IGI Global. DOI: 10.4018/978-1-5225-4769-3.ch003

Dhurpate, P. R., & Tang, H. (2021). Quantitative Analysis of the Impact of Inter-Line Conveyor Capacity for Throughput of Manufacturing Systems. *International Journal of Manufacturing, Materials, and Mechanical Engineering, 11*(1), 1–17. https://doi.org/10.4018/IJMMME.2021010101

Dinkar, S., & Deep, K. (2021). A Survey of Recent Variants and Applications of Antlion Optimizer. *International Journal of Energy Optimization and Engineering, 10*(2), 48–73. DOI: 10.4018/IJEOE.2021040103

Dixit, A. (2018). Application of Silica-Gel-Reinforced Aluminium Composite on the Piston of Internal Combustion Engine: Comparative Study of Silica-Gel-Reinforced Aluminium Composite Piston With Aluminium Alloy Piston. In Kumar, K., & Davim, J. (Eds.), *Composites and Advanced Materials for Industrial Applications* (pp. 63–98). Hershey, PA: IGI Global. DOI: 10.4018/978-1-5225-5216-1.ch004

Drabecki, M. P., & Kułak, K. B. (2021). Global Pandemics on European Electrical Energy Markets: Lessons Learned From the COVID-19 Outbreak. *International Journal of Energy Optimization and Engineering, 10*(3), 24–46. https://doi.org/10.4018/IJEOE.2021070102

Dutta, M. M. (2021). Nanomaterials for Food and Agriculture. In Bhat, M., Wani, I., & Ashraf, S. (Eds.), *Applications of Nanomaterials in Agriculture, Food Science, and Medicine* (pp. 75–97). IGI Global. DOI: 10.4018/978-1-7998-5563-7.ch004

Dutta, M. M., & Goswami, M. (2021). Coating Materials: Nano-Materials. In Roy, S., & Bose, G. (Eds.), *Advanced Surface Coating Techniques for Modern Industrial Applications* (pp. 1–30). IGI Global. DOI: 10.4018/978-1-7998-4870-7.ch001

Elsayed, A. M., Dakkama, H. J., Mahmoud, S., Al-Dadah, R., & Kaialy, W. (2017). Sustainable Cooling Research Using Activated Carbon Adsorbents and Their Environmental Impact. In Kobayashi, T. (Ed.), *Applied Environmental Materials Science for Sustainability* (pp. 186–221). Hershey, PA: IGI Global. DOI: 10.4018/978-1-5225-1971-3.ch009

Ercanoglu, M., & Sonmez, H. (2018). General Trends and New Perspectives on Landslide Mapping and Assessment Methods. In Ceryan, N. (Ed.), *Handbook of Research on Trends and Digital Advances in Engineering Geology* (pp. 350–379). Hershey, PA: IGI Global. DOI: 10.4018/978-1-5225-2709-1.ch010

Faroz, S. A., Pujari, N. N., Rastogi, R., & Ghosh, S. (2017). Risk Analysis of Structural Engineering Systems Using Bayesian Inference. In Samui, P., Chakraborty, S., & Kim, D. (Eds.), *Modeling and Simulation Techniques in Structural Engineering* (pp. 390–424). Hershey, PA: IGI Global. DOI: 10.4018/978-1-5225-0588-4.ch014

Fekik, A., Hamida, M. L., Denoun, H., Azar, A. T., Kamal, N. A., Vaidyanathan, S., Bousbaine, A., & Benamrouche, N. (2022). Multilevel Inverter for Hybrid Fuel Cell/PV Energy Conversion System. In Fekik, A., & Benamrouche, N. (Eds.), *Modeling and Control of Static Converters for Hybrid Storage Systems* (pp. 233–270). IGI Global. https://doi.org/10.4018/978-1-7998-7447-8.ch009

Fekik, A., Hamida, M. L., Houassine, H., Azar, A. T., Kamal, N. A., Denoun, H., Vaidyanathan, S., & Sambas, A. (2022). Power Quality Improvement for Grid-Connected Photovoltaic Panels Using Direct Power Control. In Fekik, A., & Benamrouche, N. (Eds.), *Modeling and Control of Static Converters for Hybrid Storage Systems* (pp. 107–142). IGI Global. https://doi.org/10.4018/978-1-7998-7447-8.ch005

Fernando, P. R., Hamigah, T., Disne, S., Wickramasingha, G. G., & Sutharshan, A. (2018). The Evaluation of Engineering Properties of Low Cost Concrete Blocks by Partial Doping of Sand with Sawdust: Low Cost Sawdust Concrete Block. *International Journal of Strategic Engineering*, *1*(2), 26–42. DOI: 10.4018/IJoSE.2018070103

Ferro, G., Minciardi, R., Parodi, L., & Robba, M. (2022). Optimal Charging Management of Microgrid-Integrated Electric Vehicles. In Alam, M., Pillai, R., & Murugesan, N. (Eds.), *Developing Charging Infrastructure and Technologies for Electric Vehicles* (pp. 133–155). IGI Global. https://doi.org/10.4018/978-1-7998-6858-3.ch007

Flumerfelt, S., & Green, C. (2022). Graduate Lean Leadership Education: A Case Study of a Program. In Alves, A., & van Hattum-Janssen, N. (Eds.), *Training Engineering Students for Modern Technological Advancement* (pp. 202–224). IGI Global. https://doi.org/10.4018/978-1-7998-8816-1.ch010

Galli, B. J. (2021). Implications of Economic Decision Making to the Project Manager. *International Journal of Strategic Engineering*, *4*(1), 19–32. https://doi.org/10.4018/IJoSE.2021010102

Gento, A. M., Pimentel, C., & Pascual, J. A. (2022). Teaching Circular Economy and Lean Management in a Learning Factory. In Alves, A., & van Hattum-Janssen, N. (Eds.), *Training Engineering Students for Modern Technological Advancement* (pp. 183–201). IGI Global. https://doi.org/10.4018/978-1-7998-8816-1.ch009

Ghosh, S., Mitra, S., Ghosh, S., & Chakraborty, S. (2017). Seismic Reliability Analysis in the Framework of Metamodelling Based Monte Carlo Simulation. In Samui, P., Chakraborty, S., & Kim, D. (Eds.), *Modeling and Simulation Techniques in Structural Engineering* (pp. 192–208). Hershey, PA: IGI Global. DOI: 10.4018/978-1-5225-0588-4.ch006

Gil, M., & Otero, B. (2017). Learning Engineering Skills through Creativity and Collaboration: A Game-Based Proposal. In Alexandre Peixoto de Queirós, R., & Pinto, M. (Eds.), *Gamification-Based E-Learning Strategies for Computer Programming Education* (pp. 14–29). Hershey, PA: IGI Global. DOI: 10.4018/978-1-5225-1034-5.ch002

Gill, J., Ayre, M., & Mills, J. (2017). Revisioning the Engineering Profession: How to Make It Happen! In Gray, M., & Thomas, K. (Eds.), *Strategies for Increasing Diversity in Engineering Majors and Careers* (pp. 156–175). Hershey, PA: IGI Global. DOI: 10.4018/978-1-5225-2212-6.ch008

Godzhaev, Z., Senkevich, S., Kuzmin, V., & Melikov, I. (2021). Use of the Neural Network Controller of Sprung Mass to Reduce Vibrations From Road Irregularities. In P. Vasant, G. Weber, & W. Punurai (Ed.), *Research Advancements in Smart Technology, Optimization, and Renewable Energy* (pp. 69-87). IGI Global. https://doi.org/10.4018/978-1-7998-3970-5.ch005

Gomes de Gusmão, C. M. (2022). Digital Competencies and Transformation in Higher Education: Upskilling With Extension Actions. In Alves, A., & van Hattum-Janssen, N. (Eds.), *Training Engineering Students for Modern Technological Advancement* (pp. 313–328). IGI Global. https://doi.org/10.4018/978-1-7998-8816-1.ch015A

Goyal, N., Ram, M., & Kumar, P. (2017). Welding Process under Fault Coverage Approach for Reliability and MTTF. In Ram, M., & Davim, J. (Eds.), *Mathematical Concepts and Applications in Mechanical Engineering and Mechatronics* (pp. 222–245). Hershey, PA: IGI Global. DOI: 10.4018/978-1-5225-1639-2.ch011

Gray, M., & Lundy, C. (2017). Engineering Study Abroad: High Impact Strategy for Increasing Access. In Gray, M., & Thomas, K. (Eds.), *Strategies for Increasing Diversity in Engineering Majors and Careers* (pp. 42–59). Hershey, PA: IGI Global. DOI: 10.4018/978-1-5225-2212-6.ch003

Güler, O., & Varol, T. (2021). Fabrication of Functionally Graded Metal and Ceramic Powders Synthesized by Electroless Deposition. In Roy, S., & Bose, G. (Eds.), *Advanced Surface Coating Techniques for Modern Industrial Applications* (pp. 150–187). IGI Global. https://doi.org/10.4018/978-1-7998-4870-7.ch007

Guraksin, G. E. (2018). Internet of Things and Nature-Inspired Intelligent Techniques for the Future of Biomedical Engineering. In Kose, U., Guraksin, G., & Deperlioglu, O. (Eds.), *Nature-Inspired Intelligent Techniques for Solving Biomedical Engineering Problems* (pp. 263–282). Hershey, PA: IGI Global. DOI: 10.4018/978-1-5225-4769-3.ch013

Hamida, M. L., Fekik, A., Denoun, H., Ardjal, A., & Bokhtache, A. A. (2022). Flying Capacitor Inverter Integration in a Renewable Energy System. In Fekik, A., & Benamrouche, N. (Eds.), *Modeling and Control of Static Converters for Hybrid Storage Systems* (pp. 287–306). IGI Global. https://doi.org/10.4018/978-1-7998-7447-8.ch011

Hasegawa, N., & Takahashi, Y. (2021). Control of Soap Bubble Ejection Robot Using Facial Expressions. *International Journal of Manufacturing, Materials, and Mechanical Engineering, 11*(2), 1–16. https://doi.org/10.4018/IJMMME.2021040101

Hejazi, T., & Akbari, L. (2017). A Multiresponse Optimization Model for Statistical Design of Processes with Discrete Variables. In Ram, M., & Davim, J. (Eds.), *Mathematical Concepts and Applications in Mechanical Engineering and Mechatronics* (pp. 17–37). Hershey, PA: IGI Global. DOI: 10.4018/978-1-5225-1639-2.ch002

Hejazi, T., & Hejazi, A. (2017). Monte Carlo Simulation for Reliability-Based Design of Automotive Complex Subsystems. In Ram, M., & Davim, J. (Eds.), *Mathematical Concepts and Applications in Mechanical Engineering and Mechatronics* (pp. 177–200). Hershey, PA: IGI Global. DOI: 10.4018/978-1-5225-1639-2.ch009

Hejazi, T., & Poursabbagh, H. (2017). Reliability Analysis of Engineering Systems: An Accelerated Life Testing for Boiler Tubes. In Ram, M., & Davim, J. (Eds.), *Mathematical Concepts and Applications in Mechanical Engineering and Mechatronics* (pp. 154–176). Hershey, PA: IGI Global. DOI: 10.4018/978-1-5225-1639-2.ch008

Henao, J., Poblano-Salas, C. A., Vargas, F., Giraldo-Betancur, A. L., Corona-Castuera, J., & Sotelo-Mazón, O. (2021). Principles and Applications of Thermal Spray Coatings. In Roy, S., & Bose, G. (Eds.), *Advanced Surface Coating Techniques for Modern Industrial Applications* (pp. 31–70). IGI Global. https://doi.org/10.4018/978-1-7998-4870-7.ch002

Henao, J., & Sotelo, O. (2018). Surface Engineering at High Temperature: Thermal Cycling and Corrosion Resistance. In Pakseresht, A. (Ed.), *Production, Properties, and Applications of High Temperature Coatings* (pp. 131–159). Hershey, PA: IGI Global. DOI: 10.4018/978-1-5225-4194-3.ch006

Hrnčič, M. K., Cör, D., & Knez, Ž. (2021). Supercritical Fluids as a Tool for Green Energy and Chemicals. In Chen, L. (Ed.), *Handbook of Research on Advancements in Supercritical Fluids Applications for Sustainable Energy Systems* (pp. 761–791). IGI Global. DOI: 10.4018/978-1-7998-5796-9.ch021

Ibrahim, O., Erdem, S., & Gurbuz, E. (2021). Studying Physical and Chemical Properties of Graphene Oxide and Reduced Graphene Oxide and Their Applications in Sustainable Building Materials. In González-Lezcano, R. (Ed.), *Advancements in Sustainable Architecture and Energy Efficiency* (pp. 221–238). IGI Global. https://doi.org/10.4018/978-1-7998-7023-4.ch010

Ihianle, I. K., Islam, S., Naeem, U., & Ebenuwa, S. H. (2021). Exploiting Patterns of Object Use for Human Activity Recognition. In Nwajana, A., & Ihianle, I. (Eds.), *Handbook of Research on 5G Networks and Advancements in Computing, Electronics, and Electrical Engineering* (pp. 382–401). IGI Global. https://doi.org/10.4018/978-1-7998-6992-4.ch015

Ijemaru, G. K., Ngharamike, E. T., Oleka, E. U., & Nwajana, A. O. (2021). An Energy-Efficient Model for Opportunistic Data Collection in IoV-Enabled SC Waste Management. In Nwajana, A., & Ihianle, I. (Eds.), *Handbook of Research on 5G Networks and Advancements in Computing, Electronics, and Electrical Engineering* (pp. 1–19). IGI Global. https://doi.org/10.4018/978-1-7998-6992-4.ch001

Ilori, O. O., Adetan, D. A., & Umoru, L. E. (2017). Effect of Cutting Parameters on the Surface Residual Stress of Face-Milled Pearlitic Ductile Iron. *International Journal of Materials Forming and Machining Processes*, *4*(1), 38–52. DOI: 10.4018/IJMFMP.2017010103

Imam, M. H., Tasadduq, I. A., Ahmad, A., Aldosari, F., & Khan, H. (2017). Automated Generation of Course Improvement Plans Using Expert System. *International Journal of Quality Assurance in Engineering and Technology Education*, *6*(1), 1–12. DOI: 10.4018/IJQAETE.2017010101

Injeti, S. K., & Kumar, T. V. (2018). A WDO Framework for Optimal Deployment of DGs and DSCs in a Radial Distribution System Under Daily Load Pattern to Improve Techno-Economic Benefits. *International Journal of Energy Optimization and Engineering*, *7*(2), 1–38. DOI: 10.4018/IJEOE.2018040101

Ishii, N., Anami, K., & Knisely, C. W. (2018). *Dynamic Stability of Hydraulic Gates and Engineering for Flood Prevention*. Hershey, PA: IGI Global. DOI: 10.4018/978-1-5225-3079-4

Iwamoto, Y., & Yamaguchi, H. (2021). Application of Supercritical Carbon Dioxide for Solar Water Heater. In Chen, L. (Ed.), *Handbook of Research on Advancements in Supercritical Fluids Applications for Sustainable Energy Systems* (pp. 370–387). IGI Global. https://doi.org/10.4018/978-1-7998-5796-9.ch010

Jayapalan, S. (2018). A Review of Chemical Treatments on Natural Fibers-Based Hybrid Composites for Engineering Applications. In Kumar, K., & Davim, J. (Eds.), *Composites and Advanced Materials for Industrial Applications* (pp. 16–37). Hershey, PA: IGI Global. DOI: 10.4018/978-1-5225-5216-1.ch002

Kapetanakis, T. N., Vardiambasis, I. O., Ioannidou, M. P., & Konstantaras, A. I. (2021). Modeling Antenna Radiation Using Artificial Intelligence Techniques: The Case of a Circular Loop Antenna. In Nikolopoulos, C. (Ed.), *Recent Trends on Electromagnetic Environmental Effects for Aeronautics and Space Applications* (pp. 186–225). IGI Global. https://doi.org/10.4018/978-1-7998-4879-0.ch007

Karkalos, N. E., Markopoulos, A. P., & Dossis, M. F. (2017). Optimal Model Parameters of Inverse Kinematics Solution of a 3R Robotic Manipulator Using ANN Models. *International Journal of Manufacturing, Materials, and Mechanical Engineering, 7*(3), 20–40. DOI: 10.4018/IJMMME.2017070102

Kelly, M., Costello, M., Nicholson, G., & O'Connor, J. (2021). The Evolving Integration of BIM Into Built Environment Programmes in a Higher Education Institute. In Underwood, J., & Shelbourn, M. (Eds.), *Handbook of Research on Driving Transformational Change in the Digital Built Environment* (pp. 294–326). IGI Global. https://doi.org/10.4018/978-1-7998-6600-8.ch012

Kesimal, A., Karaman, K., Cihangir, F., & Ercikdi, B. (2018). Excavatability Assessment of Rock Masses for Geotechnical Studies. In Ceryan, N. (Ed.), *Handbook of Research on Trends and Digital Advances in Engineering Geology* (pp. 231–256). Hershey, PA: IGI Global. DOI: 10.4018/978-1-5225-2709-1.ch006

Knoflacher, H. (2017). The Role of Engineers and Their Tools in the Transport Sector after Paradigm Change: From Assumptions and Extrapolations to Science. In Knoflacher, H., & Ocalir-Akunal, E. (Eds.), *Engineering Tools and Solutions for Sustainable Transportation Planning* (pp. 1–29). Hershey, PA: IGI Global. DOI: 10.4018/978-1-5225-2116-7.ch001

Kose, U. (2018). Towards an Intelligent Biomedical Engineering With Nature-Inspired Artificial Intelligence Techniques. In Kose, U., Guraksin, G., & Deperlioglu, O. (Eds.), *Nature-Inspired Intelligent Techniques for Solving Biomedical Engineering Problems* (pp. 1–26). Hershey, PA: IGI Global. DOI: 10.4018/978-1-5225-4769-3.ch001

Kostić, S. (2018). A Review on Enhanced Stability Analyses of Soil Slopes Using Statistical Design. In Ceryan, N. (Ed.), *Handbook of Research on Trends and Digital Advances in Engineering Geology* (pp. 446–481). Hershey, PA: IGI Global. DOI: 10.4018/978-1-5225-2709-1.ch013

Kumar, A., Patil, P. P., & Prajapati, Y. K. (2018). *Advanced Numerical Simulations in Mechanical Engineering*. Hershey, PA: IGI Global. DOI: 10.4018/978-1-5225-3722-9

Kumar, G. R., Rajyalakshmi, G., & Manupati, V. K. (2017). Surface Micro Patterning of Aluminium Reinforced Composite through Laser Peening. *International Journal of Manufacturing, Materials, and Mechanical Engineering*, *7*(4), 15–27. DOI: 10.4018/IJMMME.2017100102

Kumar, N., Basu, D. N., & Chen, L. (2021). Effect of Flow Acceleration and Buoyancy on Thermalhydraulics of sCO2 in Mini/Micro-Channel. In Chen, L. (Ed.), *Handbook of Research on Advancements in Supercritical Fluids Applications for Sustainable Energy Systems* (pp. 161–182). IGI Global. DOI: 10.4018/978-1-7998-5796-9.ch005

Kumari, N., & Kumar, K. (2018). Fabrication of Orthotic Calipers With Epoxy-Based Green Composite. In Kumar, K., & Davim, J. (Eds.), *Composites and Advanced Materials for Industrial Applications* (pp. 157–176). Hershey, PA: IGI Global. DOI: 10.4018/978-1-5225-5216-1.ch008

Kuppusamy, R. R. (2018). Development of Aerospace Composite Structures Through Vacuum-Enhanced Resin Transfer Moulding Technology (VERTMTy): Vacuum-Enhanced Resin Transfer Moulding. In Kumar, K., & Davim, J. (Eds.), *Composites and Advanced Materials for Industrial Applications* (pp. 99–111). Hershey, PA: IGI Global. DOI: 10.4018/978-1-5225-5216-1.ch005

Kurganov, V. A., Zeigarnik, Y. A., & Maslakova, I. V. (2021). Normal and Deteriorated Heat Transfer Under Heating Turbulent Supercritical Pressure Coolants Flows in Round Tubes. In Chen, L. (Ed.), *Handbook of Research on Advancements in Supercritical Fluids Applications for Sustainable Energy Systems* (pp. 494–532). IGI Global. https://doi.org/10.4018/978-1-7998-5796-9.ch014

Li, H., & Zhang, Y. (2021). Heat Transfer and Fluid Flow Modeling for Supercritical Fluids in Advanced Energy Systems. In Chen, L. (Ed.), *Handbook of Research on Advancements in Supercritical Fluids Applications for Sustainable Energy Systems* (pp. 388–422). IGI Global. https://doi.org/10.4018/978-1-7998-5796-9.ch011

Loy, J., Howell, S., & Cooper, R. (2017). Engineering Teams: Supporting Diversity in Engineering Education. In Gray, M., & Thomas, K. (Eds.), *Strategies for Increasing Diversity in Engineering Majors and Careers* (pp. 106–129). Hershey, PA: IGI Global. DOI: 10.4018/978-1-5225-2212-6.ch006

Macher, G., Armengaud, E., Kreiner, C., Brenner, E., Schmittner, C., Ma, Z., & Krammer, M. (2018). Integration of Security in the Development Lifecycle of Dependable Automotive CPS. In Druml, N., Genser, A., Krieg, A., Menghin, M., & Hoeller, A. (Eds.), *Solutions for Cyber-Physical Systems Ubiquity* (pp. 383–423). Hershey, PA: IGI Global. DOI: 10.4018/978-1-5225-2845-6.ch015

Madhu, M. N., Singh, J. G., Mohan, V., & Ongsakul, W. (2021). Transmission Risk Optimization in Interconnected Systems: Risk-Adjusted Available Transfer Capability. In P. Vasant, G. Weber, & W. Punurai (Ed.), *Research Advancements in Smart Technology, Optimization, and Renewable Energy* (pp. 183-199). IGI Global. https://doi.org/10.4018/978-1-7998-3970-5.ch010

Mahendramani, G., & Lakshmana Swamy, N. (2018). Effect of Weld Groove Area on Distortion of Butt Welded Joints in Submerged Arc Welding. *International Journal of Manufacturing, Materials, and Mechanical Engineering*, *8*(2), 33–44. DOI: 10.4018/IJMMME.2018040103

Makropoulos, G., Koumaras, H., Setaki, F., Filis, K., Lutz, T., Montowtt, P., Tomaszewski, L., Dybiec, P., & Järvet, T. (2021). 5G and Unmanned Aerial Vehicles (UAVs) Use Cases: Analysis of the Ecosystem, Architecture, and Applications. In Nwajana, A., & Ihianle, I. (Eds.), *Handbook of Research on 5G Networks and Advancements in Computing, Electronics, and Electrical Engineering* (pp. 36–69). IGI Global. https://doi.org/10.4018/978-1-7998-6992-4.ch003

Meric, E. M., Erdem, S., & Gurbuz, E. (2021). Application of Phase Change Materials in Construction Materials for Thermal Energy Storage Systems in Buildings. In González-Lezcano, R. (Ed.), *Advancements in Sustainable Architecture and Energy Efficiency* (pp. 1–20). IGI Global. https://doi.org/10.4018/978-1-7998-7023-4.ch001

Mihret, E. T., & Yitayih, K. A. (2021). Operation of VANET Communications: The Convergence of UAV System With LTE/4G and WAVE Technologies. *International Journal of Smart Vehicles and Smart Transportation*, *4*(1), 29–51. https://doi.org/10.4018/IJSVST.2021010103

Mir, M. A., Bhat, B. A., Sheikh, B. A., Rather, G. A., Mehraj, S., & Mir, W. R. (2021). Nanomedicine in Human Health Therapeutics and Drug Delivery: Nanobiotechnology and Nanobiomedicine. In Bhat, M., Wani, I., & Ashraf, S. (Eds.), *Applications of Nanomaterials in Agriculture, Food Science, and Medicine* (pp. 229–251). IGI Global. DOI: 10.4018/978-1-7998-5563-7.ch013

Mohammadzadeh, S., & Kim, Y. (2017). Nonlinear System Identification of Smart Buildings. In Samui, P., Chakraborty, S., & Kim, D. (Eds.), *Modeling and Simulation Techniques in Structural Engineering* (pp. 328–347). Hershey, PA: IGI Global. DOI: 10.4018/978-1-5225-0588-4.ch011

Molina, G. J., Aktaruzzaman, F., Soloiu, V., & Rahman, M. (2017). Design and Testing of a Jet-Impingement Instrument to Study Surface-Modification Effects by Nanofluids. *International Journal of Surface Engineering and Interdisciplinary Materials Science*, *5*(2), 43–61. DOI: 10.4018/IJSEIMS.2017070104

Moreno-Rangel, A., & Carrillo, G. (2021). Energy-Efficient Homes: A Heaven for Respiratory Illnesses. In González-Lezcano, R. (Ed.), *Advancements in Sustainable Architecture and Energy Efficiency* (pp. 49–71). IGI Global. https://doi.org/10.4018/978-1-7998-7023-4.ch003

Msomi, V., & Jantjies, B. T. (2021). Correlative Analysis Between Tensile Properties and Tool Rotational Speeds of Friction Stir Welded Similar Aluminium Alloy Joints. *International Journal of Surface Engineering and Interdisciplinary Materials Science*, *9*(2), 58–78. https://doi.org/10.4018/IJSEIMS.2021070104

Muigai, M. N., Mwema, F. M., Akinlabi, E. T., & Obiko, J. O. (2021). Surface Engineering of Materials Through Weld-Based Technologies: An Overview. In Roy, S., & Bose, G. (Eds.), *Advanced Surface Coating Techniques for Modern Industrial Applications* (pp. 247–260). IGI Global. DOI: 10.4018/978-1-7998-4870-7.ch011

Mukherjee, A., Saeed, R. A., Dutta, S., & Naskar, M. K. (2017). Fault Tracking Framework for Software-Defined Networking (SDN). In Singhal, C., & De, S. (Eds.), *Resource Allocation in Next-Generation Broadband Wireless Access Networks* (pp. 247–272). Hershey, PA: IGI Global. DOI: 10.4018/978-1-5225-2023-8.ch011

Mukhopadhyay, A., Barman, T. K., & Sahoo, P. (2018). Electroless Nickel Coatings for High Temperature Applications. In Kumar, K., & Davim, J. (Eds.), *Composites and Advanced Materials for Industrial Applications* (pp. 297–331). Hershey, PA: IGI Global. DOI: 10.4018/978-1-5225-5216-1.ch013

Mwema, F. M., & Wambua, J. M. (2022). Machining of Poly Methyl Methacrylate (PMMA) and Other Olymeric Materials: A Review. In Kumar, K., Babu, B., & Davim, J. (Eds.), *Handbook of Research on Advancements in the Processing, Characterization, and Application of Lightweight Materials* (pp. 363–379). IGI Global. https://doi.org/10.4018/978-1-7998-7864-3.ch016

Mykhailyshyn, R., Savkiv, V., Boyko, I., Prada, E., & Virgala, I. (2021). Substantiation of Parameters of Friction Elements of Bernoulli Grippers With a Cylindrical Nozzle. *International Journal of Manufacturing, Materials, and Mechanical Engineering*, *11*(2), 17–39. https://doi.org/10.4018/IJMMME.2021040102

Náprstek, J., & Fischer, C. (2017). Dynamic Stability and Post-Critical Processes of Slender Auto-Parametric Systems. In Plevris, V., Kremmyda, G., & Fahjan, Y. (Eds.), *Performance-Based Seismic Design of Concrete Structures and Infrastructures* (pp. 128–171). Hershey, PA: IGI Global. DOI: 10.4018/978-1-5225-2089-4.ch006

Nautiyal, L., Shivach, P., & Ram, M. (2018). Optimal Designs by Means of Genetic Algorithms. In Ram, M., & Davim, J. (Eds.), *Soft Computing Techniques and Applications in Mechanical Engineering* (pp. 151–161). Hershey, PA: IGI Global. DOI: 10.4018/978-1-5225-3035-0.ch007

Nazir, R. (2017). Advanced Nanomaterials for Water Engineering and Treatment: Nano-Metal Oxides and Their Nanocomposites. In Saleh, T. (Ed.), *Advanced Nanomaterials for Water Engineering, Treatment, and Hydraulics* (pp. 84–126). Hershey, PA: IGI Global. DOI: 10.4018/978-1-5225-2136-5.ch005

Nikolopoulos, C. D. (2021). Recent Advances on Measuring and Modeling ELF-Radiated Emissions for Space Applications. In Nikolopoulos, C. (Ed.), *Recent Trends on Electromagnetic Environmental Effects for Aeronautics and Space Applications* (pp. 1–38). IGI Global. https://doi.org/10.4018/978-1-7998-4879-0.ch001

Nogueira, A. F., Ribeiro, J. C., Fernández de Vega, F., & Zenha-Rela, M. A. (2018). Evolutionary Approaches to Test Data Generation for Object-Oriented Software: Overview of Techniques and Tools. In M. Khosrow-Pour, D.B.A. (Ed.), *Incorporating Nature-Inspired Paradigms in Computational Applications* (pp. 162-194). Hershey, PA: IGI Global. https://doi.org/DOI: 10.4018/978-1-5225-5020-4.ch006

Nwajana, A. O., Obi, E. R., Ijemaru, G. K., Oleka, E. U., & Anthony, D. C. (2021). Fundamentals of RF/Microwave Bandpass Filter Design. In Nwajana, A., & Ihianle, I. (Eds.), *Handbook of Research on 5G Networks and Advancements in Computing, Electronics, and Electrical Engineering* (pp. 149–164). IGI Global. https://doi.org/10.4018/978-1-7998-6992-4.ch005

Ogbodo, E. A. (2021). Comparative Study of Transmission Line Junction vs. Asynchronously Coupled Junction Diplexers. In Nwajana, A., & Ihianle, I. (Eds.), *Handbook of Research on 5G Networks and Advancements in Computing, Electronics, and Electrical Engineering* (pp. 326–336). IGI Global. https://doi.org/10.4018/978-1-7998-6992-4.ch013

Orosa, J. A., Vergara, D., Fraguela, F., & Masdías-Bonome, A. (2021). Statistical Understanding and Optimization of Building Energy Consumption and Climate Change Consequences. In González-Lezcano, R. (Ed.), *Advancements in Sustainable Architecture and Energy Efficiency* (pp. 195–220). IGI Global. https://doi.org/10.4018/978-1-7998-7023-4.ch009

Osho, M. B. (2018). Industrial Enzyme Technology: Potential Applications. In Bharati, S., & Chaurasia, P. (Eds.), *Research Advancements in Pharmaceutical, Nutritional, and Industrial Enzymology* (pp. 375–394). Hershey, PA: IGI Global. DOI: 10.4018/978-1-5225-5237-6.ch017

Ouadi, A., & Zitouni, A. (2021). Phasor Measurement Improvement Using Digital Filter in a Smart Grid. In Recioui, A., & Bentarzi, H. (Eds.), *Optimizing and Measuring Smart Grid Operation and Control* (pp. 100–117). IGI Global. https://doi.org/10.4018/978-1-7998-4027-5.ch005

Padmaja, P., & Marutheswar, G. (2017). Certain Investigation on Secured Data Transmission in Wireless Sensor Networks. *International Journal of Mobile Computing and Multimedia Communications*, *8*(1), 48–61. DOI: 10.4018/IJMCMC.2017010104

Palmer, S., & Hall, W. (2017). An Evaluation of Group Work in First-Year Engineering Design Education. In Tucker, R. (Ed.), *Collaboration and Student Engagement in Design Education* (pp. 145–168). Hershey, PA: IGI Global. DOI: 10.4018/978-1-5225-0726-0.ch007

Panchenko, V. (2021). Prospects for Energy Supply of the Arctic Zone Objects of Russia Using Frost-Resistant Solar Modules. In P. Vasant, G. Weber, & W. Punurai (Eds.), *Research Advancements in Smart Technology, Optimization, and Renewable Energy* (pp. 149-169). IGI Global. https://doi.org/10.4018/978-1-7998-3970-5.ch008

Panchenko, V. (2021). Photovoltaic Thermal Module With Paraboloid Type Solar Concentrators. *International Journal of Energy Optimization and Engineering*, *10*(2), 1–23. https://doi.org/10.4018/IJEOE.2021040101

Pandey, K., & Datta, S. (2021). Dry Machining of Inconel 825 Superalloys: Performance of Tool Inserts (Carbide, Cermet, and SiAlON). *International Journal of Manufacturing, Materials, and Mechanical Engineering*, *11*(4), 26–39. DOI: 10.4018/IJMMME.2021100102

Panneer, R. (2017). Effect of Composition of Fibers on Properties of Hybrid Composites. *International Journal of Manufacturing, Materials, and Mechanical Engineering*, *7*(4), 28–43. DOI: 10.4018/IJMMME.2017100103

Pany, C. (2021). Estimation of Correct Long-Seam Mismatch Using FEA to Compare the Measured Strain in a Non-Destructive Testing of a Pressurant Tank: A Reverse Problem. *International Journal of Smart Vehicles and Smart Transportation*, *4*(1), 16–28. DOI: 10.4018/IJSVST.2021010102

Paul, S., & Roy, P. (2018). Optimal Design of Power System Stabilizer Using a Novel Evolutionary Algorithm. *International Journal of Energy Optimization and Engineering*, *7*(3), 24–46. DOI: 10.4018/IJEOE.2018070102

Paul, S., & Roy, P. K. (2021). Oppositional Differential Search Algorithm for the Optimal Tuning of Both Single Input and Dual Input Power System Stabilizer. In P. Vasant, G. Weber, & W. Punurai (Eds.), *Research Advancements in Smart Technology, Optimization, and Renewable Energy* (pp. 256-282). IGI Global. https://doi.org/10.4018/978-1-7998-3970-5.ch013

Pavaloiu, A. (2018). Artificial Intelligence Ethics in Biomedical-Engineering-Oriented Problems. In Kose, U., Guraksin, G., & Deperlioglu, O. (Eds.), *Nature-Inspired Intelligent Techniques for Solving Biomedical Engineering Problems* (pp. 219–231). Hershey, PA: IGI Global. DOI: 10.4018/978-1-5225-4769-3.ch010

Pioro, I., Mahdi, M., & Popov, R. (2017). Application of Supercritical Pressures in Power Engineering. In Chen, L., & Iwamoto, Y. (Eds.), *Advanced Applications of Supercritical Fluids in Energy Systems* (pp. 404–457). Hershey, PA: IGI Global. DOI: 10.4018/978-1-5225-2047-4.ch013

Plaksina, T., & Gildin, E. (2017). Rigorous Integrated Evolutionary Workflow for Optimal Exploitation of Unconventional Gas Assets. *International Journal of Energy Optimization and Engineering*, *6*(1), 101–122. DOI: 10.4018/IJEOE.2017010106

Popat, J., Kakadiya, H., Tak, L., Singh, N. K., Majeed, M. A., & Mahajan, V. (2021). Reliability of Smart Grid Including Cyber Impact: A Case Study. In Singh, R., Singh, A., Dwivedi, A., & Nagabhushan, P. (Eds.), *Computational Methodologies for Electrical and Electronics Engineers* (pp. 163–174). IGI Global. https://doi.org/10.4018/978-1-7998-3327-7.ch013

Quiza, R., La Fé-Perdomo, I., Rivas, M., & Ramtahalsing, V. (2021). Triple Bottom Line-Focused Optimization of Oblique Turning Processes Based on Hybrid Modeling: A Study Case on AISI 1045 Steel Turning. In Burstein, L. (Ed.), *Handbook of Research on Advancements in Manufacturing, Materials, and Mechanical Engineering* (pp. 215–241). IGI Global. https://doi.org/10.4018/978-1-7998-4939-1.ch010

Rahmani, M. K. (2022). Blockchain Technology: Principles and Algorithms. In Khan, S., Syed, M., Hammad, R., & Bushager, A. (Eds.), *Blockchain Technology and Computational Excellence for Society 5.0* (pp. 16–27). IGI Global. https://doi.org/10.4018/978-1-7998-8382-1.ch002

Ramdani, N., & Azibi, M. (2018). Polymer Composite Materials for Microelectronics Packaging Applications: Composites for Microelectronics Packaging. In Kumar, K., & Davim, J. (Eds.), *Composites and Advanced Materials for Industrial Applications* (pp. 177–211). Hershey, PA: IGI Global. DOI: 10.4018/978-1-5225-5216-1.ch009

Ramesh, M., Garg, R., & Subrahmanyam, G. V. (2017). Investigation of Influence of Quenching and Annealing on the Plane Fracture Toughness and Brittle to Ductile Transition Temperature of the Zinc Coated Structural Steel Materials. *International Journal of Surface Engineering and Interdisciplinary Materials Science*, *5*(2), 33–42. DOI: 10.4018/IJSEIMS.2017070103

Robinson, J., & Beneroso, D. (2022). Project-Based Learning in Chemical Engineering: Curriculum and Assessment, Culture and Learning Spaces. In Alves, A., & van Hattum-Janssen, N. (Eds.), *Training Engineering Students for Modern Technological Advancement* (pp. 1–19). IGI Global. https://doi.org/10.4018/978-1-7998-8816-1.ch001

Rondon, B. (2021). Experimental Characterization of Admittance Meter With Crude Oil Emulsions. *International Journal of Electronics, Communications, and Measurement Engineering*, *10*(2), 51–59. https://doi.org/10.4018/IJECME.2021070104

Rudolf, S., Biryuk, V. V., & Volov, V. (2018). Vortex Effect, Vortex Power: Technology of Vortex Power Engineering. In Kharchenko, V., & Vasant, P. (Eds.), *Handbook of Research on Renewable Energy and Electric Resources for Sustainable Rural Development* (pp. 500–533). Hershey, PA: IGI Global. DOI: 10.4018/978-1-5225-3867-7.ch021

Sah, A., Bhadula, S. J., Dumka, A., & Rawat, S. (2018). A Software Engineering Perspective for Development of Enterprise Applications. In Elçi, A. (Ed.), *Handbook of Research on Contemporary Perspectives on Web-Based Systems* (pp. 1–23). Hershey, PA: IGI Global. DOI: 10.4018/978-1-5225-5384-7.ch001

Sahli, Y., Zitouni, B., & Hocine, B. M. (2021). Three-Dimensional Numerical Study of Overheating of Two Intermediate Temperature P-AS-SOFC Geometrical Configurations. In Badea, G., Felseghi, R., & A chilean, I. (Eds.), *Hydrogen Fuel Cell Technology for Stationary Applications* (pp. 186–222). IGI Global. https://doi.org/10.4018/978-1-7998-4945-2.ch008

Sahoo, P., & Roy, S. (2017). Tribological Behavior of Electroless Ni-P, Ni-P-W and Ni-P-Cu Coatings: A Comparison. *International Journal of Surface Engineering and Interdisciplinary Materials Science*, *5*(1), 1–15. DOI: 10.4018/IJSEIMS.2017010101

Sahoo, S. (2018). Laminated Composite Hypar Shells as Roofing Units: Static and Dynamic Behavior. In Kumar, K., & Davim, J. (Eds.), *Composites and Advanced Materials for Industrial Applications* (pp. 249–269). Hershey, PA: IGI Global. DOI: 10.4018/978-1-5225-5216-1.ch011

Sahu, H., & Hungyo, M. (2018). Introduction to SDN and NFV. In Dumka, A. (Ed.), *Innovations in Software-Defined Networking and Network Functions Virtualization* (pp. 1–25). Hershey, PA: IGI Global. DOI: 10.4018/978-1-5225-3640-6.ch001

Salem, A. M., & Shmelova, T. (2018). Intelligent Expert Decision Support Systems: Methodologies, Applications, and Challenges. In Shmelova, T., Sikirda, Y., Rizun, N., Salem, A., & Kovalyov, Y. (Eds.), *Socio-Technical Decision Support in Air Navigation Systems: Emerging Research and Opportunities* (pp. 215–242). Hershey, PA: IGI Global. DOI: 10.4018/978-1-5225-3108-1.ch007

Samal, M. (2017). FE Analysis and Experimental Investigation of Cracked and Un-Cracked Thin-Walled Tubular Components to Evaluate Mechanical and Fracture Properties. In Samui, P., Chakraborty, S., & Kim, D. (Eds.), *Modeling and Simulation Techniques in Structural Engineering* (pp. 266–293). Hershey, PA: IGI Global. DOI: 10.4018/978-1-5225-0588-4.ch009

Samal, M., & Balakrishnan, K. (2017). Experiments on a Ring Tension Setup and FE Analysis to Evaluate Transverse Mechanical Properties of Tubular Components. In Samui, P., Chakraborty, S., & Kim, D. (Eds.), *Modeling and Simulation Techniques in Structural Engineering* (pp. 91–115). Hershey, PA: IGI Global. DOI: 10.4018/978-1-5225-0588-4.ch004

Samarasinghe, D. A., & Wood, E. (2021). Innovative Digital Technologies. In Underwood, J., & Shelbourn, M. (Eds.), *Handbook of Research on Driving Transformational Change in the Digital Built Environment* (pp. 142–163). IGI Global. https://doi.org/10.4018/978-1-7998-6600-8.ch006

Sawant, S. (2018). Deep Learning and Biomedical Engineering. In Kose, U., Guraksin, G., & Deperlioglu, O. (Eds.), *Nature-Inspired Intelligent Techniques for Solving Biomedical Engineering Problems* (pp. 283–296). Hershey, PA: IGI Global. DOI: 10.4018/978-1-5225-4769-3.ch014

Schulenberg, T. (2021). Energy Conversion Using the Supercritical Steam Cycle. In Chen, L. (Ed.), *Handbook of Research on Advancements in Supercritical Fluids Applications for Sustainable Energy Systems* (pp. 659–681). IGI Global. DOI: 10.4018/978-1-7998-5796-9.ch018

Sezgin, H., & Berkalp, O. B. (2018). Textile-Reinforced Composites for the Automotive Industry. In Kumar, K., & Davim, J. (Eds.), *Composites and Advanced Materials for Industrial Applications* (pp. 129–156). Hershey, PA: IGI Global. DOI: 10.4018/978-1-5225-5216-1.ch007

Shaaban, A. A., & Shehata, O. M. (2021). Combining Response Surface Method and Metaheuristic Algorithms for Optimizing SPIF Process. *International Journal of Manufacturing, Materials, and Mechanical Engineering, 11*(4), 1–25. https://doi.org/10.4018/IJMMME.2021100101

Shafaati Shemami, M., & Sefid, M. (2022). Implementation and Demonstration of Electric Vehicle-to-Home (V2H) Application: A Case Study. In Alam, M., Pillai, R., & Murugesan, N. (Eds.), *Developing Charging Infrastructure and Technologies for Electric Vehicles* (pp. 268–293). IGI Global. https://doi.org/10.4018/978-1-7998-6858-3.ch015

Shah, M. Z., Gazder, U., Bhatti, M. S., & Hussain, M. (2018). Comparative Performance Evaluation of Effects of Modifier in Asphaltic Concrete Mix. *International Journal of Strategic Engineering*, *1*(2), 13–25. DOI: 10.4018/IJoSE.2018070102

Sharma, N., & Kumar, K. (2018). Fabrication of Porous NiTi Alloy Using Organic Binders. In Kumar, K., & Davim, J. (Eds.), *Composites and Advanced Materials for Industrial Applications* (pp. 38–62). Hershey, PA: IGI Global. DOI: 10.4018/978-1-5225-5216-1.ch003

Shivach, P., Nautiyal, L., & Ram, M. (2018). Applying Multi-Objective Optimization Algorithms to Mechanical Engineering. In Ram, M., & Davim, J. (Eds.), *Soft Computing Techniques and Applications in Mechanical Engineering* (pp. 287–301). Hershey, PA: IGI Global. DOI: 10.4018/978-1-5225-3035-0.ch014

Shmelova, T. (2018). Stochastic Methods for Estimation and Problem Solving in Engineering: Stochastic Methods of Decision Making in Aviation. In Kadry, S. (Ed.), *Stochastic Methods for Estimation and Problem Solving in Engineering* (pp. 139–160). Hershey, PA: IGI Global. DOI: 10.4018/978-1-5225-5045-7.ch006

Siero González, L. R., & Romo Vázquez, A. (2017). Didactic Sequences Teaching Mathematics for Engineers With Focus on Differential Equations. In Ramírez-Montoya, M. (Ed.), *Handbook of Research on Driving STEM Learning With Educational Technologies* (pp. 129–151). Hershey, PA: IGI Global. DOI: 10.4018/978-1-5225-2026-9.ch007

Sim, M. S., You, K. Y., Esa, F., & Chan, Y. L. (2021). Nanostructured Electromagnetic Metamaterials for Sensing Applications. In Bhat, M., Wani, I., & Ashraf, S. (Eds.), *Applications of Nanomaterials in Agriculture, Food Science, and Medicine* (pp. 141–164). IGI Global. https://doi.org/10.4018/978-1-7998-5563-7.ch009

Singh, R., & Dutta, S. (2018). Visible Light Active Nanocomposites for Photocatalytic Applications. In Kumar, K., & Davim, J. (Eds.), *Composites and Advanced Materials for Industrial Applications* (pp. 270–296). Hershey, PA: IGI Global. DOI: 10.4018/978-1-5225-5216-1.ch012

Skripov, P. V., Yampol'skiy, A. D., & Rutin, S. B. (2021). High-Power Heat Transfer in Supercritical Fluids: Microscale Times and Sizes. In Chen, L. (Ed.), *Handbook of Research on Advancements in Supercritical Fluids Applications for Sustainable Energy Systems* (pp. 424–450). IGI Global. https://doi.org/10.4018/978-1-7998-5796-9.ch012

Sözbilir, H., Özkaymak, Ç., Uzel, B., & Sümer, Ö. (2018). Criteria for Surface Rupture Microzonation of Active Faults for Earthquake Hazards in Urban Areas. In Ceryan, N. (Ed.), *Handbook of Research on Trends and Digital Advances in Engineering Geology* (pp. 187–230). Hershey, PA: IGI Global. DOI: 10.4018/978-1-5225-2709-1.ch005

Stanciu, I. (2018). Stochastic Methods in Microsystems Engineering. In Kadry, S. (Ed.), *Stochastic Methods for Estimation and Problem Solving in Engineering* (pp. 161–176). Hershey, PA: IGI Global. DOI: 10.4018/978-1-5225-5045-7.ch007

Strebkov, D., Nekrasov, A., Trubnikov, V., & Nekrasov, A. (2018). Single-Wire Resonant Electric Power Systems for Renewable-Based Electric Grid. In Kharchenko, V., & Vasant, P. (Eds.), *Handbook of Research on Renewable Energy and Electric Resources for Sustainable Rural Development* (pp. 449–474). Hershey, PA: IGI Global. DOI: 10.4018/978-1-5225-3867-7.ch019

Sukhyy, K., Belyanovskaya, E., & Sukhyy, M. (2021). *Basic Principles for Substantiation of Working Pair Choice*. IGI Global. DOI: 10.4018/978-1-7998-4432-7.ch002

Suri, M. S., & Kaliyaperumal, D. (2022). Extension of Aspiration Level Model for Optimal Planning of Fast Charging Stations. In Fekik, A., & Benamrouche, N. (Eds.), *Modeling and Control of Static Converters for Hybrid Storage Systems* (pp. 91–106). IGI Global. https://doi.org/10.4018/978-1-7998-7447-8.ch004

Tallet, E., Gledson, B., Rogage, K., Thompson, A., & Wiggett, D. (2021). Digitally-Enabled Design Management. In Underwood, J., & Shelbourn, M. (Eds.), *Handbook of Research on Driving Transformational Change in the Digital Built Environment* (pp. 63–89). IGI Global. https://doi.org/10.4018/978-1-7998-6600-8.ch003

Terki, A., & Boubertakh, H. (2021). A New Hybrid Binary-Real Coded Cuckoo Search and Tabu Search Algorithm for Solving the Unit-Commitment Problem. *International Journal of Energy Optimization and Engineering, 10*(2), 104–119. https://doi.org/10.4018/IJEOE.2021040105

Tüdeş, Ş., Kumlu, K. B., & Ceryan, S. (2018). Integration Between Urban Planning and Natural Hazards For Resilient City. In Ceryan, N. (Ed.), *Handbook of Research on Trends and Digital Advances in Engineering Geology* (pp. 591–630). Hershey, PA: IGI Global. DOI: 10.4018/978-1-5225-2709-1.ch017

Ulamis, K. (2018). Soil Liquefaction Assessment by Anisotropic Cyclic Triaxial Test. In Ceryan, N. (Ed.), *Handbook of Research on Trends and Digital Advances in Engineering Geology* (pp. 631–664). Hershey, PA: IGI Global. DOI: 10.4018/978-1-5225-2709-1.ch018

Valente, M., & Milani, G. (2017). Seismic Assessment and Retrofitting of an Under-Designed RC Frame Through a Displacement-Based Approach. In Plevris, V., Kremmyda, G., & Fahjan, Y. (Eds.), *Performance-Based Seismic Design of Concrete Structures and Infrastructures* (pp. 36–58). Hershey, PA: IGI Global. DOI: 10.4018/978-1-5225-2089-4.ch002

Vargas-Bernal, R. (2021). Advances in Electromagnetic Environmental Shielding for Aeronautics and Space Applications. In Nikolopoulos, C. (Ed.), *Recent Trends on Electromagnetic Environmental Effects for Aeronautics and Space Applications* (pp. 80–96). IGI Global. https://doi.org/10.4018/978-1-7998-4879-0.ch003

Vasant, P. (2018). A General Medical Diagnosis System Formed by Artificial Neural Networks and Swarm Intelligence Techniques. In Kose, U., Guraksin, G., & Deperlioglu, O. (Eds.), *Nature-Inspired Intelligent Techniques for Solving Biomedical Engineering Problems* (pp. 130–145). Hershey, PA: IGI Global. DOI: 10.4018/978-1-5225-4769-3.ch006

Verner, C. M., & Sarwar, D. (2021). Avoiding Project Failure and Achieving Project Success in NHS IT System Projects in the United Kingdom. *International Journal of Strategic Engineering*, *4*(1), 33–54. https://doi.org/10.4018/IJoSE.2021010103

Verrollot, J., Tolonen, A., Harkonen, J., & Haapasalo, H. J. (2018). Challenges and Enablers for Rapid Product Development. *International Journal of Applied Industrial Engineering*, *5*(1), 25–49. DOI: 10.4018/IJAIE.2018010102

Wan, A. C., Zulu, S. L., & Khosrow-Shahi, F. (2021). Industry Views on BIM for Site Safety in Hong Kong. In Underwood, J., & Shelbourn, M. (Eds.), *Handbook of Research on Driving Transformational Change in the Digital Built Environment* (pp. 120–140). IGI Global. https://doi.org/10.4018/978-1-7998-6600-8.ch005

Yardimci, A. G., & Karpuz, C. (2018). Fuzzy Rock Mass Rating: Soft-Computing-Aided Preliminary Stability Analysis of Weak Rock Slopes. In Ceryan, N. (Ed.), *Handbook of Research on Trends and Digital Advances in Engineering Geology* (pp. 97–131). Hershey, PA: IGI Global. DOI: 10.4018/978-1-5225-2709-1.ch003

You, K. Y. (2021). Development Electronic Design Automation for RF/Microwave Antenna Using MATLAB GUI. In Nwajana, A., & Ihianle, I. (Eds.), *Handbook of Research on 5G Networks and Advancements in Computing, Electronics, and Electrical Engineering* (pp. 70–148). IGI Global. https://doi.org/10.4018/978-1-7998-6992-4.ch004

Yousefi, Y., Gratton, P., & Sarwar, D. (2021). Investigating the Opportunities to Improve the Thermal Performance of a Case Study Building in London. *International Journal of Strategic Engineering*, *4*(1), 1–18. https://doi.org/10.4018/IJoSE.2021010101

Zindani, D., & Kumar, K. (2018). Industrial Applications of Polymer Composite Materials. In Kumar, K., & Davim, J. (Eds.), *Composites and Advanced Materials for Industrial Applications* (pp. 1–15). Hershey, PA: IGI Global. DOI: 10.4018/978-1-5225-5216-1.ch001

Zindani, D., Maity, S. R., & Bhowmik, S. (2018). A Decision-Making Approach for Material Selection of Polymeric Composite Bumper Beam. In Kumar, K., & Davim, J. (Eds.), *Composites and Advanced Materials for Industrial Applications* (pp. 112–128). Hershey, PA: IGI Global. DOI: 10.4018/978-1-5225-5216-1.ch006

About the Contributors

Shelli Brunswick, a renowned figure in space innovation and leadership. With a diverse career from the U.S. Air Force to COO of Space Foundation, and now CEO and Founder of SB Global. She champions space technology innovation, entrepreneurship, and diversity. As a prolific author and speaker, she's gives talks globally and contributes to publications like SpaceNews and Forbes Technology Council. Notably, Shelli shapes conversations on innovation and technology through engagements with think tanks like the Hudson Institute and research for NASA and ESA. Recognized with accolades including Top 100 Women of the Future in Emerging Technology, she's dedicated to empowering women in space and fosters global collaboration through international organizations like UNOOSA Space4Women Mentoring Program and WomenTech Network. Shelli Brunswick is a trailblazer propelling us towards an expansive future in space exploration and leadership.

Kyla Tennin is an industry practitioner and DM in Management and Organizational Leadership.She focuses on Change Management and mixed methods empirical research for continuous improvement, business transformation, and recovery during and after internal/external crisis to nations, institutions, and individual's lives. She is a leader in various roles at investment forums where she works with Presidents of Nations, Heads of Banks, Prime Ministers, HighCommissioners, and Chambers of Commerce and other business leaders to strategize and make businessdecisions to help businesses start-up, advance, and/or scale to improve economic development, promote social justice, and create jobs in various nations. At the World Business Angels Investment Forum(WBAF), affiliated partner of the G20 Global Partnership for Financial Inclusion (GPFI), this vision is to promote social justice, create jobs, financial inclusion, and innovative financial instruments for innovators, startups, and SMEs, helping them to scale. Dr. Tennin is also affiliated with the International Monetary Fund, World Bank Group, and Parliamentary Network to enhance economies and various forms of businesses.(Contact info:

Tilottama Singh is a certified HR analyst from IIM and a proficient academic, researcher, and trainer with more than 16 years of experience in the field of Human Resources and Work Dynamics. Presently working as Head of Deparment in School of Management, Uttaranchal Institute of Management, Uttaranchal University, Dehradun. Her areas of expertise are Human Resources & Strategy, and her teaching concentration includes Human Resources, Strategy, and Law. She has obtained her Master's in Human Resources Management with Distinction. She also serves as a member of AIMA and acts to lead and liaise between the student community and the industry delegates, with a keen interest in training and building sustainable models for business and society.

Richa Goel is an associate professor-economics and International Business at SCMS Noida, Symbiosis International Deemed University, Pune. She has a Ph.D. in Management, where she has worked for almost 6 years in the area of Diversity Management. She has almost 20+ years of experience in academics along with authoring more than 50+ research papers for publication in reputed national and international journals accompanied by hundreds of papers for international/national conferences and symposiums. She has written more than 25 book chapters for leading publishers, and she has served as a member of the review committee for conferences and journals. She is also currently a series editor for two different series for Taylor and Francis.

***

Harry is a faculty member in the Department of Sociology at Lovely Professional University, India. Her expertise lies in migration studies, where she has made significant contributions through research papers and book chapters published in reputed journals and edited volumes. Beyond her academic pursuits, Dr. Harry has actively organized numerous cultural events, fostering student engagement and interdisciplinary dialogue. As an active member of the department, she plays a vital role in promoting academic excellence and cultural awareness.

Sovannaroth Chheang earned her bachelor's degree in English from Build Bright University, Phnom Penh, in 2016, where she built a solid foundation in language studies, communication, and pedagogy. She began her teaching career in 2013 as a Tutor at the Kindergarten Association, conducting both part-time and full-time English programs. Since 2018, she has served as a Teacher of English and General Assistant at Sovanna Elton Foreign Language School. Through these roles, she has developed broad expertise in classroom management, language instruction, curriculum design, and student engagement, supporting the school's mission of advancing foreign language learning. Ms. Chheang's academic interests center on pursuing a master's degree in English education or related fields within the humanities and social sciences. Passionate about integrating global research methodologies into her practice, she aims to enhance pedagogy, foster classroom innovation, and promote effective, research-informed teaching. Her commitment lies in bridging academic research with real-world practice to advance English education in Cambodia and beyond.

Chheangkhy Hok is a lecturer at the Department of Khmer Literature, University of Heng Samrin Thbongkhmum. With a strong dedication to the study and preservation of Khmer language, literature, and culture, he has contributed to advancing literary education and research in Cambodia. His academic interests include classical and modern Khmer literature, linguistic analysis, and cultural studies. Mr. Chheangkhy is passionate about inspiring students to explore Cambodia's rich literary heritage and promoting the value of Khmer identity through education. In addition to teaching, he actively engages in academic writing and community cultural initiatives, aiming to strengthen the role of literature in contemporary Cambodian society.

Sovanna Huot is a PhD candidate in Political Science at the University of Delhi, New Delhi, India. He is a vice principal at a Government Public High School of Cambodia and is responsible for the school academic and institutional leadership. He is a very proficient academic across many disciplines. He has completed a two-year teacher training program from 2010-2012 and has obtained bachelor's degrees in English and in Educational Administration. He also has two master's degrees in law and in TEFL obtained through a very competitive scholarship, and a Master of Education, specialized in School Leadership, Upper Secondary School Principalship, from the National Institute of Education (NIE), Phnom Penh, 2021. He specializes in educational reform and in the practice-policy nexus of digit pedagogy and in glonacal (global–national–local) education systems. He has formal academic training in political science and education and has published scholarly work in both disciplines, including articles indexed in Scopus (Author ID: 59316244400).

Sovanna Loch is a prominent figure at the Cambodian Monastery Rajgir, serving as a deputy in charge. He is also a PhD research scholar and ICCR Fellow in the Dept. of English at Nava Nalanda Mahavihara (Deemed University), an institution under the Ministry of Culture, Govt. of India. His academic background includes a bachelor's degree in Pāli and a master's in English, both obtained from NNM. His research primarily focuses on Cambodian classical and Buddhist English literature.

Kimyut Phin is an educator and PhD research scholar in English at the University of Delhi, India. He serves as the director of the Institute of Foreign Languages of the University of Heng Samrin Thbongkhmum, Cambodia. Kimyut has also had teaching experiences in linguistics, applied linguistics, teaching methodology, second language acquisition, academic writing skills, global studies, and literature studies for years. Further, he has experience in lesson planning, classroom management, assessment and evaluation, and technology for teaching and learning. His research centers on ELT and linguistics, education, digital pedagogy, literary and theoretical studies, cultural and gender studies, and translation studies, and his work also gains Scopus-indexed publications (Scopus Author ID: 60516584900).

Kritika Sharma is a Research Scholar in Sociology at Lovely Professional University, India. Her research interests lie in tribal studies, cultural studies, gender studies, and ethnography, with a particular focus on qualitative methods such as thematic analysis. She has published several research papers and book chapters in reputed journals and edited volumes, contributing to the growing discourse on indigenous knowledge systems and social structures. In addition to her scholarly contributions, she actively serves as a reviewer for academic journals, reflecting her commitment to advancing research in her field.

Index

W

www.ingramcontent.com/pod-product-compliance
Lightning Source LLC
LaVergne TN
LVHW062225200926
848484LV00006B/1478